PRACTICE MAKES PERFECT

Geometry

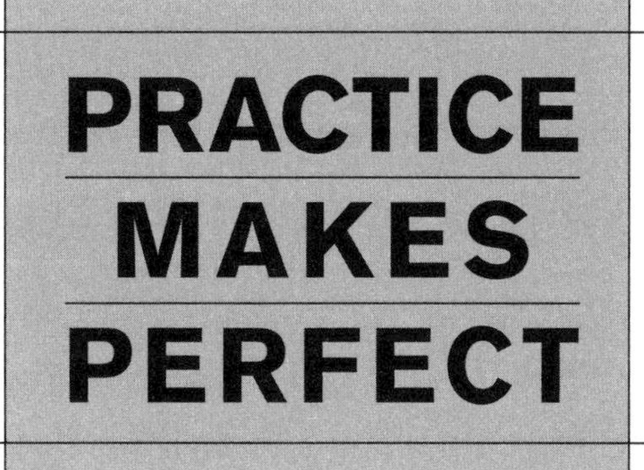

Geometry

Carolyn Wheater

New York Chicago San Francisco Lisbon London Madrid Mexico City
Milan New Delhi San Juan Seoul Singapore Sydney Toronto

The McGraw-Hill Companies

1 2 3 4 5 6 7 8 9 10 11 12 13 14 15 WDQ/WDQ 1 9 8 7 6 5 4 3 2 1 0

ISBN 978-0-07-163814-2
MHID 0-07-163814-8

Library of Congress Control Number: 2009941938

Interior design by Village Bookworks, Inc.
Interior illustrations by Glyph International

This book is printed on acid-free paper.

Contents

Introduction

An old joke tells of a tourist, lost in New York City, who stops a passerby to ask, "How do I get to Carnegie Hall?" The New Yorker's answer comes back quickly: "Practice, practice, practice!" The joke may be lame, but it contains a truth. No musician performs on the stage of a renowned concert hall without years of daily and diligent practice. No dancer steps out on stage without hours in the rehearsal hall, and no athlete takes to the field or the court without investing time and sweat drilling on the skills of his or her sport.

Math has a lot in common with music, dance, and sports. There are skills to be learned and a sequence of activities you need to go through if you want to be good at it. You don't just read math, or just listen to math, or even just understand math. You *do* math, and to learn to do it well, you have to practice. That's why homework exists, but most people need more practice than homework provides. That's where *Practice Makes Perfect: Geometry* comes in.

Many students of geometry are intimidated by the list of postulates and theorems they are expected to learn, but there are a few important things you should remember about acquiring that knowledge. First, it doesn't all happen at once. No learning ever does, but in geometry in particular, postulates and theorems are introduced a few at a time, and they build on the ones that have come before. The exercises in this book are designed to take you through each of those ideas, step-by-step. The key to remembering all that information is not, as many people think, rote memorization. A certain amount of memory work is necessary, but what solidifies the ideas in your mind and your memory is using them, putting them to work. The exercises in *Practice Makes Perfect: Geometry* let you put each idea into practice by solving problems based on each principle.

You'll find numerical problems and problems that require you to use skills from algebra. You'll also see questions that ask you to draw conclusions or to plan a proof. As you work, take the time to draw a diagram (if one is not provided) and mark up the diagram to indicate what you know or can easily conclude. Write out your work clearly enough that you can read it back to find and correct errors. Use the answers provided at the end of the book to check your work.

With patience and practice, you'll find that you've not only learned all your postulates and theorems but also come to understand why they are true and how they fit together. More than any fact or group of facts, that ability to understand the logical system is the most important thing you will take away from geometry. It will serve you well in other math courses and in other disciplines. Be persistent. You must keep working at it, bit by bit. Be patient. You will make mistakes, but mistakes are one of the ways we learn, so welcome your mistakes. They'll decrease as you practice, because practice makes perfect.

Logic and reasoning

·1·

One thing that distinguishes a course in geometry from the bits and pieces of geometric information you've picked up over the years is a greater emphasis on logical reasoning. You'll be asked to make conjectures based on evidence and to prove those conjectures using deductive reasoning.

Patterns

Observing patterns that occur in your work can help you formulate a conjecture, an educated guess about what can be expected to happen. If you drew several rectangles and measured their diagonals, you might predict that the diagonals of a rectangle are always the same length. That isn't proof, but it's a place to begin.

When you're searching for patterns, ask yourself what the items you're looking at have in common, but consider too what changes from one number or object to the next. You want to know what changes as well as what stays the same. Take a moment to look for a counterexample, a case that shows that your conjecture is wrong. Finding even one example that contradicts your hypothesis is enough to show it's not true, but examples alone can't prove it's always true.

EXERCISE 1·1

Give the next item in the pattern.

1. 3, 7, 11, 15, . . .
2. $\frac{1}{2}, \frac{3}{4}, \frac{5}{6}, \frac{7}{8}, \ldots$
3.

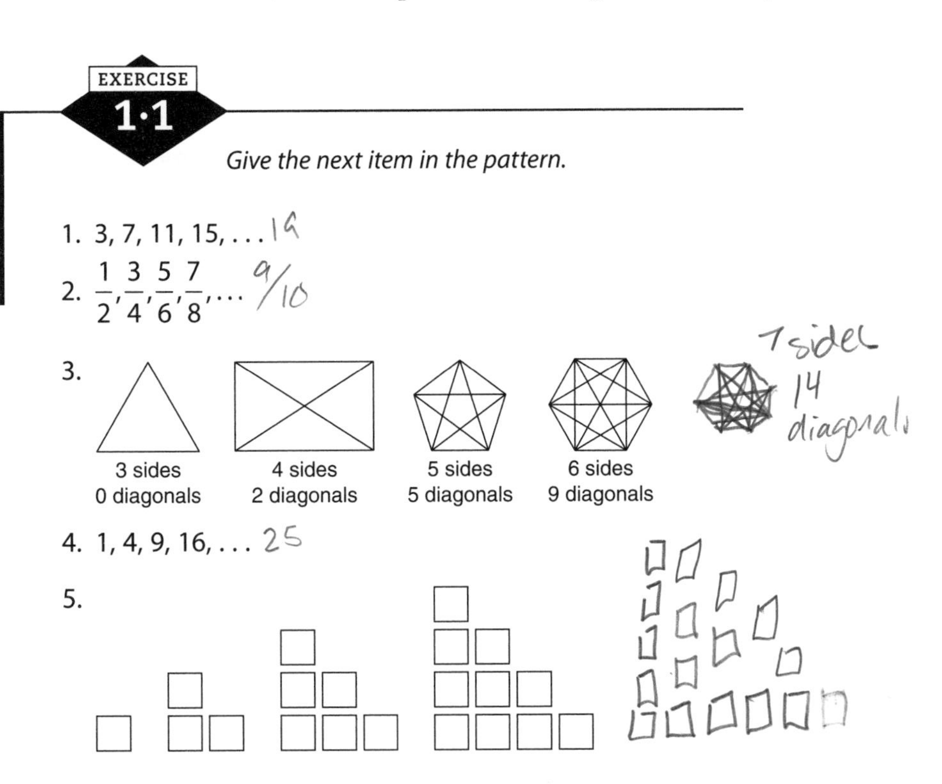

4. 1, 4, 9, 16, . . .
5.

Make a conjecture based on the information given. What evidence would disprove the conjecture?

6. $1^2 = 1$, $3^2 = 9$, $5^2 = 25$, $7^2 = 49$, $9^2 = 81$
7. You draw several rectangles and measure the diagonals. For each rectangle, the two diagonals are the same length.
8. The last three times you wore a green shirt, your English teacher gave a pop quiz.
9. You draw several triangles and measure the sides. In each triangle, the lengths of the two shorter sides total more than the length of the longest side.
10. 1, $2 = (1) + 1$, $4 = (2 + 1) + 1$, $8 = (4 + 2 + 1) + 1$, $16 = (8 + 4 + 2 + 1) + 1$

Statements

A *statement* is a sentence that is either true or false. Its truth can be determined. Sentences such as "Albany is the capital of New York" and "$7 = 5 + 3$" are examples of statements. The first one is true, and the second is false. Open sentences are sentences that contain variables. It may be a variable in the sense of a symbol that stands for a number, as in "$x + 8 = 12$" and "$4 - \square = 1$," or it may be a word whose meaning can't be determined. The sentence "It is red" is an open sentence. You can't label it true or false until you know what "it" is. Because the value of the variable is not known, the truth of the sentence cannot be determined.

When you study logic, you focus on statements, because you're interested in whether they are true or false. We usually use a single lowercase letter, such as p or q, to stand for a statement. If we talk about the truth value of a statement, we're just talking about whether it's true or false. Every statement is either true or false. If it's a statement, there's no other option.

Negation

The *negation* of a statement is its logical opposite. The negation is usually formed by placing the word *not* in the sentence. If p is the statement "The math test is Friday," then the negation of p, which we write $\sim p$, is the statement "The math test is not Friday." Be careful when you form a negation. "The math test is Thursday" doesn't say the same thing as "The math test is not Friday."

If a statement is true, its negation is false. If a statement is false, its negation is true. A statement and its negation have opposite truth values. You can summarize this in what's called a *truth table.*

p	$\sim p$
T	F
F	T

Two wrongs don't make a right, but two negations do get you back where you started from, or in symbolic form, $\sim(\sim p)$ is p.

NEGATION

p	$\sim p$	$\sim(\sim p)$
T	F	T
F	T	F

Conjunction

A *conjunction* is an *and* statement. It's a new statement formed by joining two statements with the word *and*. The symbol for AND is $\wedge$. If p is the statement "My dog is brown" and q is the statement "My cat is white," then $p \wedge q$ is the statement "My dog is brown and my cat is white."

Is that true? Well, no, actually my dog is black and my cat is gray. What if I had a black dog and a white cat? Would $p \wedge q$ be true? No, it wouldn't be true to say, "My dog is brown and my cat is white." Only part of that statement is true, and that's not good enough. If you connect the pieces with the word *and*, then both pieces must be true in order for the new statement—the conjunction—to be true. A conjunction is only true if both of the statements that form it are true.

CONJUNCTION

p	q	$p \wedge q$
T	T	T
T	F	F
F	T	F
F	F	F

Disjunction

If I offered you cake and ice cream, you would expect to get both. On the other hand, if I offered you cake or ice cream, you'd probably expect to get only one. Now it's possible that I'd end up giving you some of each, but I wouldn't be obligated to give you both. I said cake *or* ice cream, not cake *and* ice cream.

A *disjunction* is an OR statement. A new statement formed by connecting two statements with the word *or* is called a disjunction. The symbol for OR is $\vee$. If p is the statement "I will go to the dentist" and q is the statement "I will go shopping," then the disjunction $p \vee q$ is the statement "I will go to the dentist or I will go shopping." If I go to the dentist, my statement is true. If I go shopping, my statement is true. If I go shopping before visiting the dentist, or if I feel well enough after the dentist to stop at the mall, my statement is still true. The only way that the statement "I will go to the dentist or I will go shopping" would not be true is if I did neither thing. If I didn't go to the dentist and didn't go shopping, then my statement would not be true. A disjunction is only false if both of the statements that form it are false.

DISJUNCTION

p	q	$p \vee q$
T	T	T
T	F	T
F	T	T
F	F	F

Use statements p, q, r, and t to write out each of the symbolic statements below, and tell whether the new statement is true or false.

p: A touchdown scores 6 points. (True)

q: A field goal scores 3 points. (True)

r: You can kick for an extra point after a touchdown. (True)

t: A fumble scores 10 points. (False)

1. $p \wedge r$
2. $q \vee t$
3. $\sim r \vee p$
4. $q \wedge \sim t$
5. $\sim p \vee \sim q$
6. $t \wedge \sim p$
7. $\sim p \wedge \sim t$
8. $q \vee \sim r$
9. $\sim r \wedge \sim t$
10. $p \vee q \vee \sim r$

Complete each of the following truth tables.

11.

p	q	$\sim p$	$\sim p \vee q$
T	T		
T	F		
F	T		
F	F		

12.

p	q	$\sim p$	$\sim q$	$\sim p \wedge \sim q$
T	T			
T	F			
F	T			
F	F			

13.

p	q	$\sim q$	$p \wedge \sim q$
T	T		
T	F		
F	T		
F	F		

14.

p	q	$p \wedge q$	$\sim(p \wedge q)$
T	T		
T	F		
F	T		
F	F		

15.

p	q	$\sim p$	$\sim q$	$\sim p \vee \sim q$
T	T			
T	F			
F	T			
F	F			

Conditionals and biconditionals

A *conditional* statement is an "If . . . , then . . ." statement. The "if" clause is called the *hypothesis* or *premise*, and the clause that follows the word *then* is the *conclusion*. The symbol for the conditional is $\rightarrow$. If the hypothesis is denoted by p and the conclusion by q, then the conditional "if p, then q" can be symbolized as $p \rightarrow q$.

The truth of a conditional is tricky. Think about the conditional as a promise. The promise "If you get an A, then I'll stand on my head" can only be broken if you do get an A and I don't stand on my head. If you get the A, you've met the condition. If I stand on my head, I've kept my promise and the conditional is true. If you've met the condition by getting the A, but I refuse to stand on my head, I've broken my promise and the conditional is false. If you get any other grade, you haven't met my condition. I'm not obligated to do anything. I can stay firmly on my feet, and I haven't broken my promise. My statement is true. But I can also decide to stand on my head, just because I want to, and I haven't lied to you. My promise was about what I'd do if you got an A. You didn't, so there's no way I can break the promise.

		CONDITIONAL
p	q	$p \to q$
T	T	T
T	F	F
F	T	T
F	F	T

A conditional can be false only if the hypothesis is true and the conclusion is false. A conditional statement is false only if a true premise leads to a false conclusion.

Related conditionals

Every conditional statement has three related conditionals, called the *converse*, the *inverse*, and the *contrapositive*. Many of the logical errors people make stem from confusion about similar-sounding conditional statements, so it's important to keep them straight.

Converse

The *converse* of $p \to q$ is $q \to p$. The converse swaps the hypothesis and the conclusion. The converse of "If you get an A, then I'll stand on my head" is the statement "If I stand on my head, then you get an A." You can probably hear that they don't say the same thing, but people often confuse them.

When both p and q are true, or both p and q are false, then both the original conditional and its converse are true. In any other situation, however, the truth values are not the same.

		CONDITIONAL	CONVERSE
p	q	$p \to q$	$q \to p$
T	T	T	T
T	F	F	T
F	T	T	F
F	F	T	T

Inverse

The *inverse* of a conditional negates each statement in the conditional. The inverse of "If you get an A, then I'll stand on my head" is "If you don't get an A, then I won't stand on my head." The traditional symbol for the negative or opposite of p is $\sim p$, so the inverse can be symbolized as $\sim p \to \sim q$. The truth of the conditional does not guarantee the truth of the inverse, just as it doesn't guarantee the truth of the converse; but the converse and the inverse always have the same truth value; that is, they're either both true or both false.

				CONDITIONAL	CONVERSE	INVERSE
p	q	$\sim p$	$\sim q$	$p \to q$	$q \to p$	$\sim p \to \sim q$
T	T	F	F	T	T	T
T	F	F	T	F	T	T
F	T	T	F	T	F	F
F	F	T	T	T	T	T

Contrapositive

There is a related conditional that is true whenever $p \to q$ is true, however. The *contrapositive* $\sim q \to \sim p$ is equivalent to $p \to q$. The contrapositive of "If you get an A, then I'll stand on my head" is "If I don't stand on my head, then you didn't get an A."

In the truth table below, notice that the column for the conditional and the column for the contrapositive are identical.

				CONDITIONAL	CONVERSE	INVERSE	CONTRAPOSITIVE
p	q	$\sim p$	$\sim q$	$p \to q$	$q \to p$	$\sim p \to \sim q$	$\sim q \to \sim p$
T	T	F	F	T	T	T	T
T	F	F	T	F	T	T	F
F	T	T	F	T	F	F	T
F	F	T	T	T	T	T	T

The fact that a conditional statement is true does not guarantee that its converse or its inverse will be true. There are cases when a statement and its converse are both true and other cases when they're not. The contrapositive, however, is true whenever the original is true.

Biconditional

The *biconditional* is the conjunction of two conditionals. A biconditional statement is actually a conditional and its converse, but it's often condensed to avoid repetitive language. It is usually written "p if and only if q" but it means "If p, then q, and if q, then p." The symbol for the biconditional is $\leftrightarrow$.

p	q	$p \to q$	$q \to p$	$(p \to q) \wedge (q \to p)$	$p \leftrightarrow q$
T	T	T	T	T	T
T	F	F	T	F	F
F	T	T	F	F	F
F	F	T	T	T	T

The biconditional "If the weather is damp, then there will be many mosquitoes, and if there are many mosquitoes, then the weather is damp" can be condensed to "There are many mosquitoes if and only if the weather is damp." Symbolically, you write a biconditional as $p \leftrightarrow q$, since it's $(p \to q) \wedge (q \to p)$. Because a biconditional is essentially saying that p and q are equivalent statements, it's true when both p and q are true or both are false.

Write the converse, inverse, and contrapositive of each conditional.

1. If wood is burned, then it turns to ash.
2. If a team scores a touchdown, then they kick an extra point.
3. If you don't study for your exam, then you don't score well.
4. If you don't eat lunch, then you'll be hungry in the afternoon.
5. If you run a marathon, then you'll be tired.

Form the biconditional (in condensed form) and tell whether it is true or false.

6. A number is odd. A number is 1 less than an even number.
7. A number is divisible by 3. A number is divisible by 6.
8. A number is a multiple of 3. The digits of a number add to a multiple of 3.
9. A triangle is equilateral. A triangle is equiangular.
10. The first day of the month is a Tuesday. The 30th day of the month is a Thursday.

Complete each truth table.

11.

p	q	$\sim p$	$\sim p \rightarrow q$
T	T		
T	F		
F	T		
F	F		

12.

p	q	$\sim p$	$\sim q$	$\sim p \wedge \sim q$	$(\sim p \wedge \sim q) \rightarrow q$
T	T				
T	F				
F	T				
F	F				

13.

p	q	$\sim q$	$p \wedge \sim q$	$(p \wedge \sim q) \rightarrow p$
T	T			
T	F			
F	T			
F	F			

14.

p	q	$\sim p$	$\sim q$	$p \wedge q$	$\sim(p \wedge q)$	$[\sim(p \wedge q)] \leftrightarrow (\sim p \vee \sim q)$
T	T					
T	F					
F	T					
F	F					

15.

p	q	$\sim p$	$\sim q$	$p \vee q$	$\sim(p \vee q)$	$\sim p \wedge \sim q$	$[\sim(p \vee q)] \leftrightarrow (\sim p \wedge \sim q)$
T	T						
T	F						
F	T						
F	F						

Deduction

Deductive reasoning begins with a premise or hypothesis and reasons to a conclusion according to clear rules. People use deductive reasoning all the time, but all too often, they think their reasoning is valid when it's not. There are a lot of common logical errors, some that come from thinking a statement says something it doesn't, and others that result from trying to link statements together when they really don't link up.

Valid arguments

An argument, or pattern of reasoning, is valid if true assumptions always lead to true conclusions. Knowing and following patterns of logical argument are your best plan for constructing a clear proof. The two most common patterns of reasoning are known as *detachment* and *syllogism*.

Detachment

If you know that a conditional statement $p \to q$ is true and that p is true, then you can conclude that q is true. "If you get an A, then I'll stand on my head" and "You get an A" allow you to conclude, "I'll stand on my head."

When you construct a truth table and the last column is always true, you have a tautology. Every valid pattern of reasoning results in a tautology. Here's the truth table for detachment.

p	q	$p \to q$	$(p \to q) \wedge p$	$[(p \to q) \wedge p] \to q$
T	T	T	T	T
T	F	F	F	T
F	T	T	F	T
F	F	T	F	T

Syllogism

Syllogism is probably the most powerful tool in reasoning, because it lets you chain together several statements. If you know that $p \to q$ is true and that $q \to r$ is true, then you can conclude that $p \to r$ is true. "If the temperature is over 90 degrees, then we'll run the air conditioner" and "If we run the air conditioner, then our electric bill will go up" allow you to conclude that "If the temperature is over 90 degrees, then our electric bill will go up."

The truth table for syllogism is much larger because it involves p, q, and r. Instead of the four possibilities you need to consider for statements involving p and q, when you work with three statements, there are eight possibilities.

p	q	r	$p \to q$	$q \to r$	$p \to r$	$(p \to q) \wedge (q \to r)$	$[(p \to q) \wedge (q \to r)] \to (p \to r)$
T	T	T	T	T	T	T	T
T	T	F	T	F	F	F	T
T	F	T	F	T	T	F	T
T	F	F	F	T	F	F	T
F	T	T	T	T	T	T	T
F	T	F	T	F	T	F	T
F	F	T	T	T	T	T	T
F	F	F	T	T	T	T	T

Logical errors

Common errors in logic include reasoning from the converse or the inverse. These occur when we try to draw conclusions based on the converse or inverse of the given statement, although we can't be sure the converse and inverse are true. "If the ladder tips, then I fall" and "I fell" do not guarantee that the ladder tipped, although people will often incorrectly reach that conclusion.

Constructing a truth table for the pattern of argument used in reasoning from the converse or reasoning from the inverse will show you that the patterns are not valid, because the truth tables will not result in tautologies. Here's the truth table for reasoning from the converse.

p	q	$q \rightarrow p$	$(q \rightarrow p) \wedge p$	$[(q \rightarrow p) \wedge p] \rightarrow q$
T	T	T	T	T
T	F	T	T	F
F	T	F	F	T
F	F	T	F	T

The truth table for reasoning from the inverse has a similar flaw. It's not a tautology either, so neither reasoning from the converse nor reasoning from the inverse is a valid argument.

p	q	$\sim p$	$\sim q$	$\sim p \rightarrow \sim q$	$(\sim p \rightarrow \sim q) \wedge p$	$[(\sim p \rightarrow \sim q) \wedge p] \rightarrow q$
T	T	F	F	T	T	T
T	F	F	T	T	T	F
F	T	T	F	F	F	T
F	F	T	T	T	F	T

Indirect proof

It's OK to reason from the contrapositive, however, because the contrapositive is true whenever the original is true. Reasoning from the contrapositive is sometimes called *proof by contradiction* or *indirect proof*. The conditional "If school is closed, then we will go to a movie" has a contrapositive "If we don't go to a movie, then school is not closed." Since those are logically equivalent statements, you can substitute one for the other in your reasoning.

p	q	$\sim p$	$\sim q$	$\sim q \rightarrow \sim p$	$(\sim q \rightarrow \sim p) \wedge p$	$[(\sim q \rightarrow \sim p) \wedge p] \rightarrow q$
T	T	F	F	T	T	T
T	F	F	T	F	F	T
F	T	T	F	T	F	T
F	F	T	T	T	F	T

As the truth table shows, reasoning from the contrapositive is a tautology, and therefore a valid pattern of reasoning. But what does an indirect argument sound like? Suppose you know that the statement "If you step out of bounds, then the play is over" is a true statement, and you know the play is not over. You can logically conclude that you didn't step out of bounds, because the fact that "If you step out of bounds, then the play is over" is true means that its contrapositive, "If the play is not over, then you did not step out of bounds," is also true.

EXERCISE
1·4

State the conclusion that can be drawn based on the given statements, or write "No conclusion." Assume the given statements are true.

1. If the fire alarm rings, then everyone leaves the building. The fire alarm is ringing.
2. If you study geometry, then you improve your logic skills. If you improve your logic skills, then your essays will be better structured.
3. If I eat a big lunch, I get sleepy in the afternoon. I took an afternoon nap.
4. If you practice an instrument every day, your playing will improve. You may have a career in music if your playing improves.
5. A car will not run if there is no gas in the tank. There is gas in the tank.

Tell whether the conclusion drawn is valid. If not, explain the error.

6. If you want to become a doctor, then you should study science. You do not want to become a doctor; therefore, you do not need to study science.
7. If the milk is sour, it will curdle in your coffee. If the milk curdles in your coffee, then you will not drink it. Therefore, if the milk is sour, you will not drink your coffee.
8. If you practice an instrument every day, you will improve your playing. If you want a career in music, you will improve your playing. Therefore, if you want a career in music, you will practice an instrument every day.
9. If you do your math homework, you will pass the math test. You'll be able to play video games if you do your math homework. Therefore if you play video games, you'll pass the math test.
10. If it rains, then you carry an umbrella. You are not carrying an umbrella; therefore, it is not raining.

Reasoning in algebra

In algebra, you learned about properties of arithmetic, and you applied them to solving equations. The reasoning you did as you solved equations is similar to the reasoning you'll do in geometry. Each step follows logically from the one before and is justified by a property you've learned. In algebra, you used properties like these:

Associative property	$(a+b)+c=a+(b+c)$, $(ab)c=a(bc)$
Commutative property	$a+b=b+a$, $ab=ba$
Distributive property	$a(b+c)=ab+ac$
Addition property of equality	If $a=b$, then $a+c=b+c$
Subtraction property of equality	If $a=b$, then $a-c=b-c$
Multiplication property of equality	If $a=b$, then $ac=bc$
Division property of equality	If $a=b$ and $c \neq 0$, then $\frac{a}{c}=\frac{b}{c}$
Substitution property	If $a=b$, then a can be replaced with b.

EXERCISE 1·5

Identify the property used in each example.

1. $$3x-7=32$$ $$3x-7+7=32+7$$ $$3x=39$$

2. $$3x=1$$ $$\frac{3x}{3}=\frac{1}{3}$$ $$x=\frac{1}{3}$$

3. $$\frac{x}{2}=8$$ $$\frac{x}{2}\cdot 2=8\cdot 2$$ $$x=16$$

4. $$11x+18-3x=3x-14-3x$$ $$11x-3x+18=3x-3x-14$$ $$8x+18=-14$$

5. $5x=(x+12)-8$

$5x=x+(12-8)$

$5x=x+4$

6. $8x-17=12+3x$

$8x-17-3x=12+3x-3x$

$8x-3x-17=12+3x-3x$

$5x-17=12$

7. $2x+5y=24$

$y=2x$

$2x+5(2x)=24$

8. $4(x-7)+6=18$

$4x-28+6=18$

Solve each equation, justifying each step.

9. $8(x-4)-16=10(x-7)$

10. $8(2x-5)-2(x-2)=5(x+7)-4(x+8)$

Constructing an argument

Many times in geometry you are asked to prove that a statement is true. Some people do this in a paragraph form, and others organize the steps in their thinking in one column and the reasons in another. In geometry, the reasons you'll give are definitions, postulates, and theorems.

Postulates and theorems

A statement we accept as true without proof is called a *postulate*. Postulates are usually statements whose truth is clear from experience. Fundamental postulates include these:

- Two points determine one and only one line.
- Three points not all on the same line determine one and only one plane.
- A line contains at least two points.
- Given a line in a plane, there exists a point in the plane not on that line.
- Every line is a set of points that can be put into a one-to-one correspondence with real numbers.
- On a number line, there is a unique distance between two points.
- If two points lie on a plane, the line containing them also lies on the plane.
- If two planes intersect, their intersection is a line.

A *theorem* is a statement that is proved true by using postulates and theorems already proven. Once a theorem is proved, it can be used to prove others.

Construct an argument to prove the statement is true, or give a counterexample.

1. For a line and a point not on that line, there is exactly one plane that contains them.
2. If two lines intersect, they intersect in no more than one point.
3. Two intersecting lines contain at least three points.
4. If two lines intersect, there is exactly one plane that contains them.
5. If two lines do not intersect, there is exactly one plane that contains them.

6. If a plane contains two lines that do not intersect, then that plane contains at least four points.
7. If A, B, and C are all on the same line, then the distance from A to C plus the distance from C to B is equal to the distance from A to B.
8. If A, B, and C are distinct points and the distance from A to C plus the distance from C to B is greater than the distance from A to B, then A, B, and C do not all lie on the same line.
9. If a plane contains a line, it contains at least two lines.
10. If two planes have a point in common, then they have at least two points in common.

Lines and angles

·2·

The building blocks of geometry are points, lines, and planes. Such fundamental ideas are sometimes difficult or impossible to define, and since we have to start somewhere, we take those terms as undefined. Then we start to combine points and lines into more complex figures.

Undefined terms

The words *point*, *line*, and *plane* are described but not formally defined.

A *point* is usually represented by a dot but has no dimension. It takes up no space, not even the space occupied by a tiny dot. It has a location, but no length, no width, and no thickness. A point is usually labeled with a capital letter.

A *line* has infinite length, like a string of points going on forever. Lines have no width or thickness. A line can be named by a single lowercase letter or by placing a double-pointed arrow over two capital letters that name points on the line.

- Every line contains at least two points.
- Through any two points, there is one and only one line.
- Two points that lie on the same line are collinear points.

You often hear people talk about "a straight line" but all lines are straight. If two lines intersect, they intersect in exactly one point. They cannot bend to cross again.

A *plane* is a flat surface, going on forever. It has infinite length and infinite width, but no thickness. A plane is usually named by a single capital letter.

Three or more points, not all collinear, define a plane. Since a line contains at least two points, a line and a point not on the line give you at least three points, and so a line and a point not on the line also define a plane.

If two planes intersect, their intersection is a line.

EXERCISE

2·1

Use the following diagram to answer these questions.

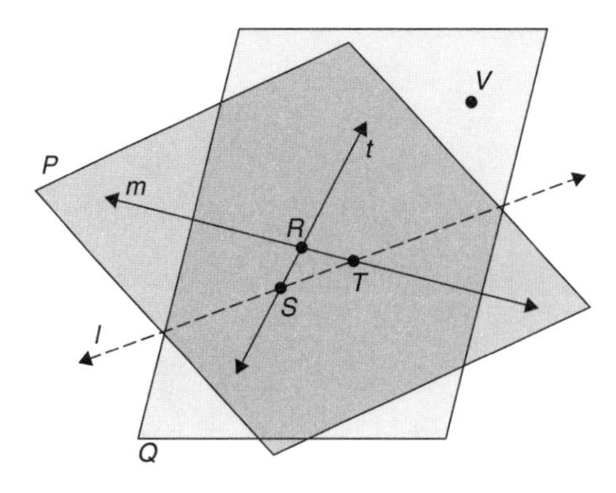

1. Name two points on line *l*.
2. Name a line that passes through point *R*.
3. Name a plane that contains point *T*.
4. Name a line that lies on plane *P*.
5. Name the plane that contains line *m*.
6. Name the intersection of line *l* and line *m*.
7. Name the intersection of plane *P* and plane *Q*.
8. Name a line that intersects line *t*.
9. Name a line that does not lie in plane *Q*.
10. Name a plane that contains line *l* and point *V*.

Rays and segments

A *ray* is a portion of a line from a point, called the *endpoint*, and continuing without end in one direction. A ray is named by the endpoint and another point that the ray passes through, with an arrow over the top. $\overrightarrow{AB}$ names a ray that starts from A and runs through B and beyond, while $\overrightarrow{BA}$ is a ray that starts at B and runs through A and beyond. (See Figure 2.1.)

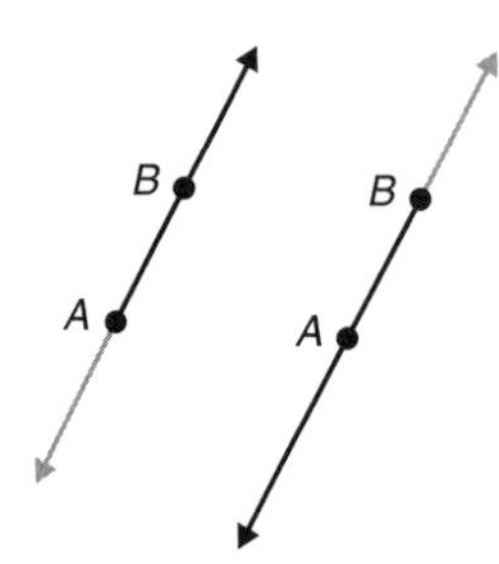

Figure 2.1 A ray is a portion of a line.

Two rays with the same endpoint that form a line are called *opposite rays*.

A *segment* is a portion of a line between two endpoints. A segment is named by its endpoints with a segment drawn over the top. The segment with endpoints A and B can be named as $\overline{AB}$ or $\overline{BA}$.

Use the following diagram to answer these questions.

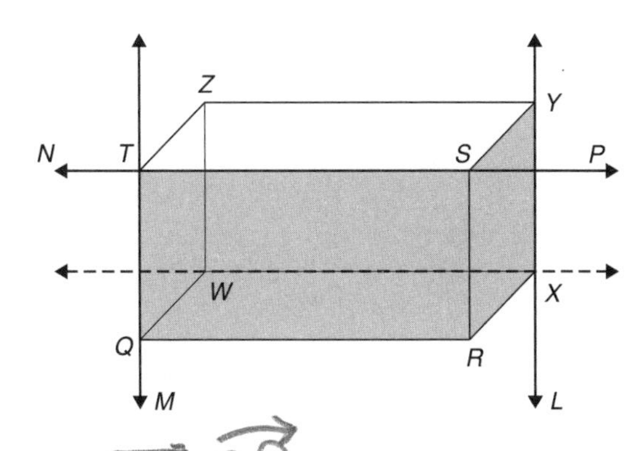

1. Name a ray with endpoint S.
2. Name a pair of opposite rays with endpoint X.
3. Name a ray that with $\overrightarrow{SN}$ forms a pair of opposite rays.
4. Name a pair of rays with endpoint T that are not opposite rays.
5. Name a ray that passes through T but does not have T as its endpoint.
6. True or false? $\overrightarrow{TS}$ and $\overrightarrow{SN}$ are opposite rays.
7. True or false? $\overrightarrow{YX}$ and $\overrightarrow{YL}$ name the same ray.
8. Name a segment that has X as an endpoint.
9. Name two different segments that have Q as an endpoint.
10. True or false? $\overline{SR}$ and $\overline{RS}$ name the same segment.

Measuring lengths

A line goes on forever, so it has infinite length. It is not possible to measure a line, to put a number on it, because it has no beginning and no end. While a ray has one endpoint, it goes on forever in the other direction, so it cannot be measured either. Segments, because they have two endpoints, have a finite length. A segment can be assigned a number that tells how long it is. The number that describes the length of a segment will depend on the units of measurement (inches, centimeters, etc.) that are used.

The ruler postulate says that it is possible to assign numbers to points on a line so that the length of a segment can be found by subtracting the numbers assigned to the endpoints. The number that corresponds to a point is called its *coordinate*. In Figure 2.2, the coordinate of point D is 2. Point A has a coordinate of –7. In geometry, distance is always positive (or more accurately, the distance between two points is the same regardless of direction), so the distance between two points is the absolute value of the difference of their coordinates. The distance between point B and point E is $|-2-5| = 7$ units.

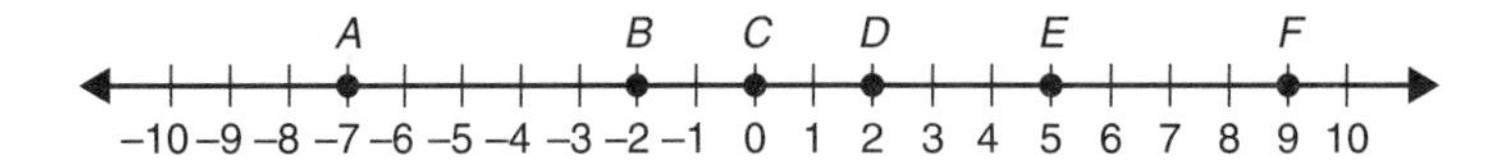

Figure 2.2 Points on a line and their coordinates.

The length of a segment is the distance between its endpoints, so the length of $\overline{EF}$ is the distance between E and F, which is $|5-9| = 4$ units. The length of $\overline{AB}$ is $|-7-(-2)| = 5$. Write AB, without any symbol over the top, to mean the length of segment $\overline{AB}$. The length of $\overline{BE}$ is $BE = 7$ units, and $AB = 5$ units is the length of $\overline{AB}$.

Two segments that have the same length are congruent. If $AB = CD$, then $\overline{AB} \cong \overline{CD}$.

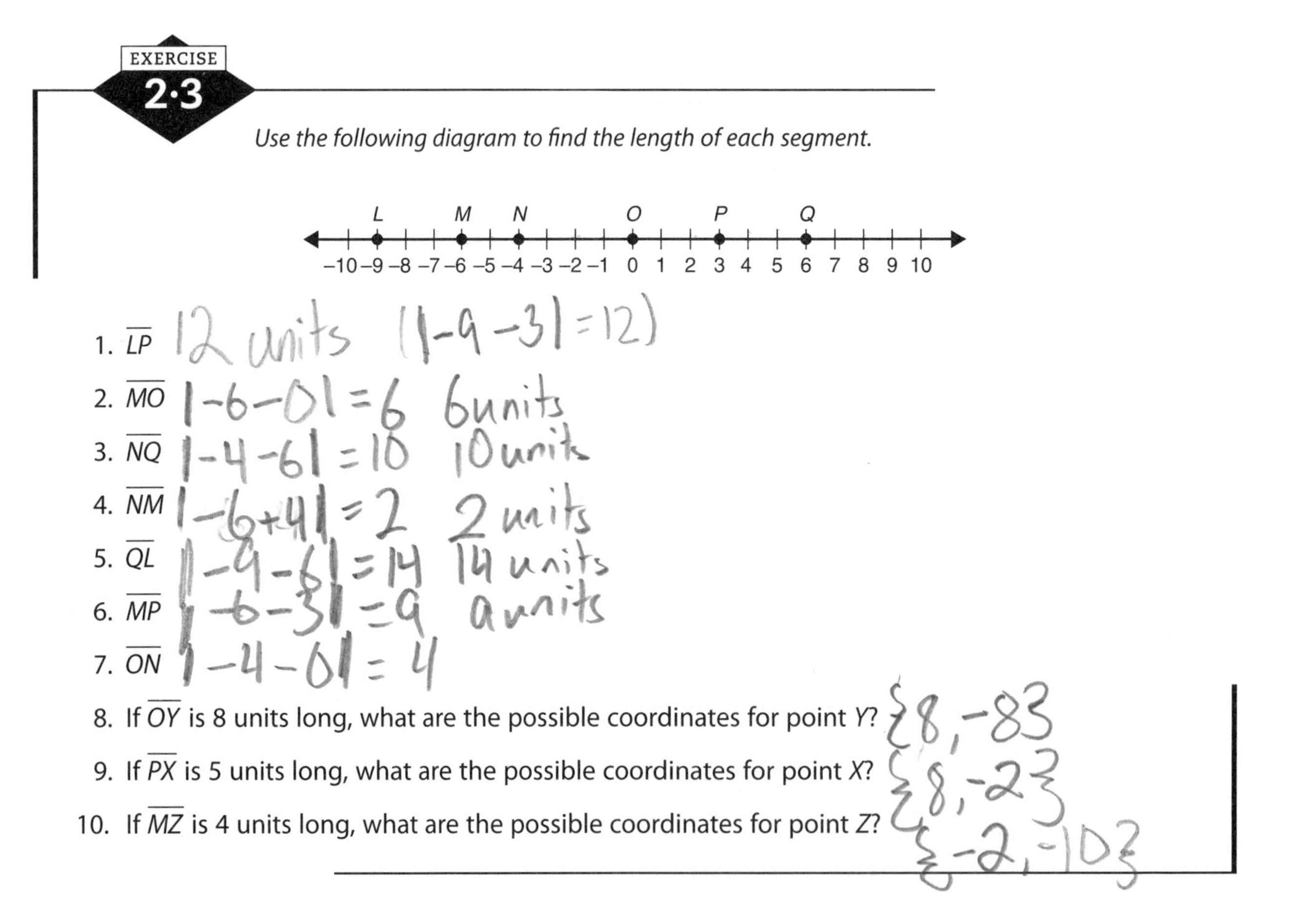

EXERCISE 2·3

Use the following diagram to find the length of each segment.

1. $\overline{LP}$
2. $\overline{MO}$
3. $\overline{NQ}$
4. $\overline{NM}$
5. $\overline{QL}$
6. $\overline{MP}$
7. $\overline{ON}$
8. If $\overline{OY}$ is 8 units long, what are the possible coordinates for point Y?
9. If $\overline{PX}$ is 5 units long, what are the possible coordinates for point X?
10. If $\overline{MZ}$ is 4 units long, what are the possible coordinates for point Z?

Angles

An *angle* is formed by two rays with a common endpoint. The common endpoint is called the *vertex*, and the two rays are the *sides* of the angle. In Figure 2.3, the point Y is the vertex of the angle, and rays $\overrightarrow{YX}$ and $\overrightarrow{YZ}$ are the sides.

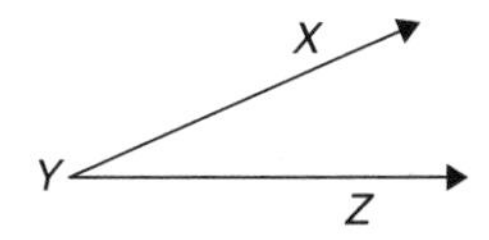

Figure 2.3

You will sometimes see angles that are formed by two segments even though this does not exactly fit the definition of an angle. (See Figure 2.4.) Just imagine that the segments are parts of rays.

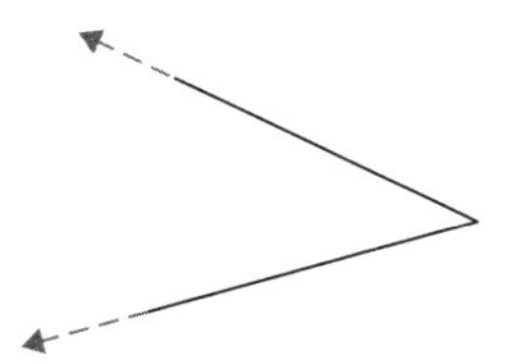

Figure 2.4

Angles are named by naming a point on one side, then the vertex, then a point on the other side. The three letters outline the angle. The angle in Figure 2.3 can be named $\angle XYZ$ or $\angle ZYX$. The vertex letter must be in the middle.

If there is only one angle with a particular vertex and no confusion is possible, the angle can be named by just its vertex letter. The angle in Figure 2.3 can be named $\angle Y$, because it is the only angle with vertex Y, but in Figure 2.5, $\angle P$ is not a useful name because there are three angles with vertex P: $\angle RPS$, $\angle QPS$, and $\angle RPQ$.

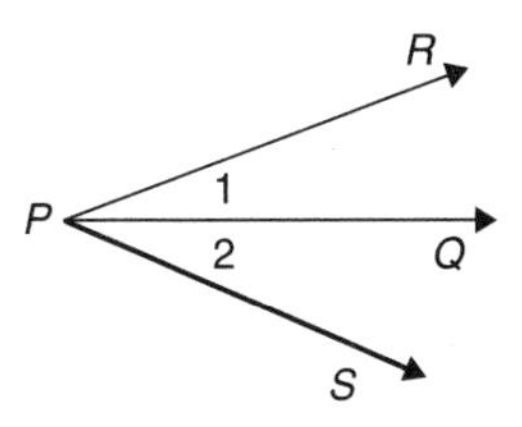

Figure 2.5

When there are several different angles at the same vertex, it can be helpful to number the angles. Thus $\angle RPQ$ can be named as $\angle 1$ and $\angle QPS$ as $\angle 2$.

Use the following diagram to answer these questions.

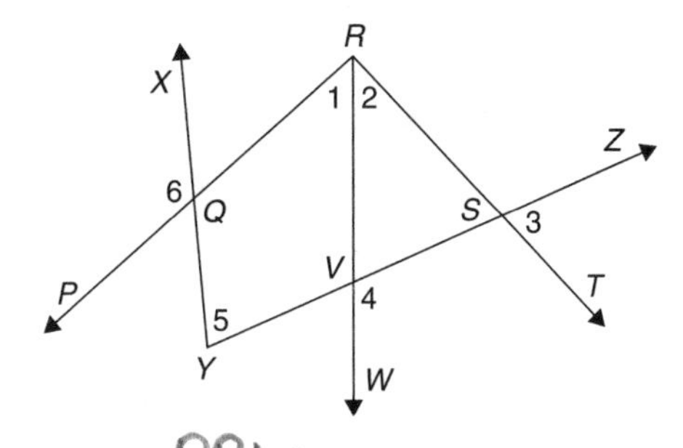

1. Name $\angle 1$ with three letters. $\angle PRW$
2. Name $\angle 4$ with three letters. $\angle WVZ$

3. Name three angles with vertex at R.
4. Give another name for $\angle 6$.
5. Give another name for $\angle TSZ$.
6. What is the vertex of $\angle 5$?
7. Name the sides of $\angle 2$.
8. True or false? $\angle WVS$ is another name for $\angle WVZ$.
9. True or false? $\angle YQV$ is another name for $\angle 5$.
10. True or false? $\angle RVY$ is another name for $\angle 4$.

Measuring angles

The measurement of an angle has nothing to do with how long the sides are. The size of the angle is a measurement of the rotation or separation between the sides. In geometry, angles are measured in degrees. A full rotation, all the way around a circle, is 360 degrees (360°). If the sides of the angle are rotated so that they point in opposite directions and form a pair of opposite rays, the measure of the angle is 180°. The hands of a clock make a 180° angle at 6:00. At 6:15, they form only a quarter rotation, a 90° angle.

The *protractor postulate* says that all the rays from a single point can be assigned numbers, or coordinates, so that when the coordinates of the two sides of an angle are subtracted, the absolute value of the difference is the measure of the angle. The instrument called a protractor is a semicircle with markings for the degrees from 0 to 180. Most protractors have two scales: one goes clockwise and the other counterclockwise. Be certain you read the coordinates for both sides of the angle on the same scale.

On the top scale of the protractor in Figure 2.6, the coordinate of $\overrightarrow{XY}$ is 140 and the coordinate of $\overrightarrow{XZ}$ is 50. The measure of $\angle ZXY = |140 - 50| = 90°$. The coordinate of $\overrightarrow{XW}$ is 85, so the measure of $\angle ZXW$ is, written $m\angle ZXW = |50 - 85| = 35°$. To find the measure of $\angle WXY$, subtract the coordinates of $\overrightarrow{XW}$ and $\overrightarrow{XY}$ and take the absolute value. So $\angle WXY$ measures $|85 - 140| = 55°$.

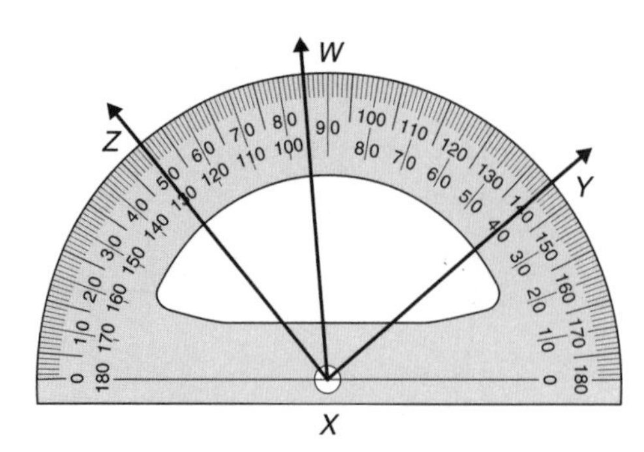

Figure 2.6

We write a small m in front of the angle name to mean "the measure of" the angle. If two angles have the same measurement, then they are congruent. And so $\angle XYZ \cong \angle ABC$ means that $m\angle XYZ = m\angle ABC$.

Classifying angles

Angles can be classified according to their measure. Angles that measure less than 90° are called *acute* angles. An angle that measures exactly 90° is a *right* angle. Two lines (or rays or segments) that form a right angle are *perpendicular.* If an angle measures more than 90° but less than 180°, the angle is an *obtuse* angle. Opposite rays form an angle of 180°, called a *straight* angle, because it looks like a straight line.

Use the following diagram to find the measure of each angle.

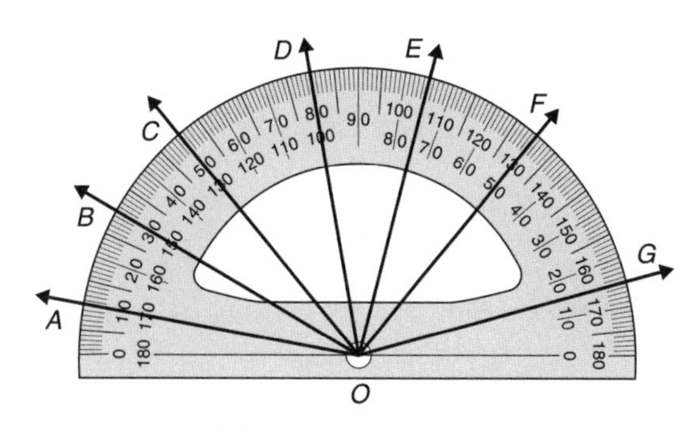

1. ∠AOF |10-130| = 120°
2. ∠BOE |30-105| = 75°
3. ∠COD |50-80| = 30°
4. ∠DOB 80-30 = 50°
5. ∠EOG |105-165| = 60°
6. ∠FOB 130 − 30 = 100°
7. ∠GOC 165 − 50 = 115°
8. ∠EOA 105 − 10 = 95°
9. ∠COB 50 − 30 = 20°
10. ∠DOG |80 − 165| = 85°

Use the following diagram to classify each angle as acute, right, obtuse, *or* straight.

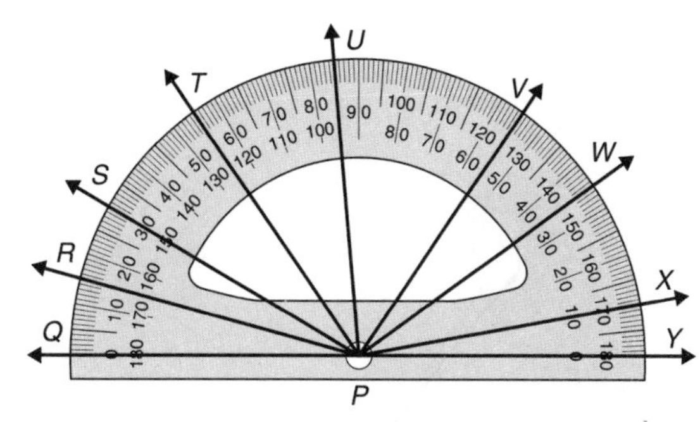

11. ∠RPS acute
12. ∠QPY straight
13. ∠UPW acute
14. ∠TPW right
15. ∠QPR acute
16. ∠UPX acute
17. ∠RPV obtuse
18. ∠QPX obtuse
19. ∠WPX acute
20. ∠XPR obtuse

Midpoints and bisectors

The *midpoint* of a segment is a point on the segment that is the same distance from one endpoint as it is from the other. Point M is the midpoint of $\overline{AB}$ if M is a point on $\overline{AB}$ and $AM = MB$. Each line segment has one and only one midpoint. A *bisector of a segment* is any line, ray, or segment that passes through the midpoint of the segment. A segment has exactly one midpoint, but it can have many bisectors.

If the bisector of a segment also makes a right angle with the segment, it is called the *perpendicular bisector.* Every point on the perpendicular bisector of a segment is equidistant from the endpoints of the segment.

The *bisector of an angle* is a ray that has the vertex of the angle as its endpoint and falls between the sides of the angle so that it divides the angle into two angles of equal size. $\overrightarrow{AB}$ is the bisector of $\angle CAD$ if $\overrightarrow{AB}$ is between $\overrightarrow{AC}$ and $\overrightarrow{AD}$ and $m\angle CAB = m\angle BAD$. An angle has exactly one bisector.

Use the following diagram to answer questions 1–5.

1. What is the midpoint of $\overline{OQ}$? P
2. If A is the midpoint of $\overline{NO}$, what is the coordinate of A? −2
3. If B is the midpoint of $\overline{MN}$, what is the coordinate of B? −5
4. If M is the midpoint of $\overline{LC}$, what is the coordinate of C? −3
5. If P is the midpoint of $\overline{ND}$, what is the coordinate of D? 10

Use the following diagram to answer questions 6–10.

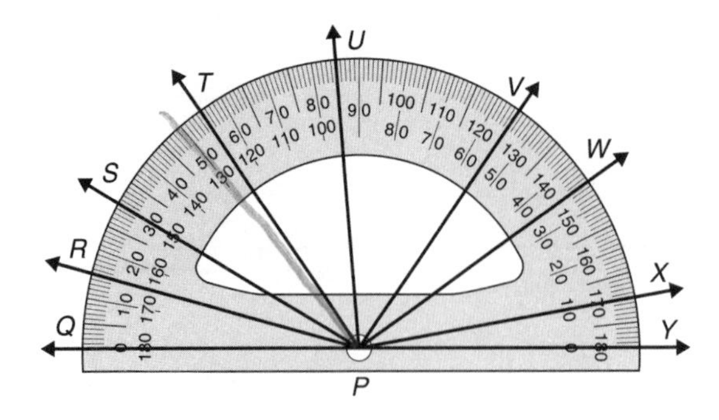

6. True or false? Since $m\angle TPU = m\angle WPX$, $\overrightarrow{PU}$ is the bisector of $\angle TPX$.

7. True or false? $\overrightarrow{PT}$ is the bisector of $\angle SPU$.
8. Name the bisector of $\angle QPS$.
9. Give the coordinate of the bisector of $\angle TPW$.
10. Give the coordinate of the bisector of $\angle RPU$.

Angle relationships

There are several different relationships of angle pairs that frequently occur in geometry problems.

Complements and supplements

Two angles whose measurements add to 90° are *complementary* angles. Each angle is the *complement* of the other. Two angles whose measurements add to 180° are *supplementary* angles. Each angle is the *supplement* of the other. Keep the names straight by remembering that 90 comes before 180 numerically and complementary comes before supplementary alphabetically.

Vertical angles

When two lines intersect, the X-shaped figure they create has two pairs of vertical angles. *Vertical angles* have a common vertex, but no shared sides. Their sides form two pairs of opposite rays. The two angles in a pair of vertical angles are always congruent.

Linear pairs

Two angles that have a common vertex, share a side, and do not overlap are *adjacent angles.* When adjacent angles have exterior sides that form a line (or a straight angle), they are called a *linear pair.* The two angles in a linear pair are always supplementary.

EXERCISE **2·7**

Use the following diagram to answer questions 5–10.

1. Find the measure of the complement of an angle of 53°.
2. Find the measure of the complement of an angle of 31°.
3. Find the measure of the supplement of an angle of 47°.
4. Find the measure of the supplement of an angle of 101°.

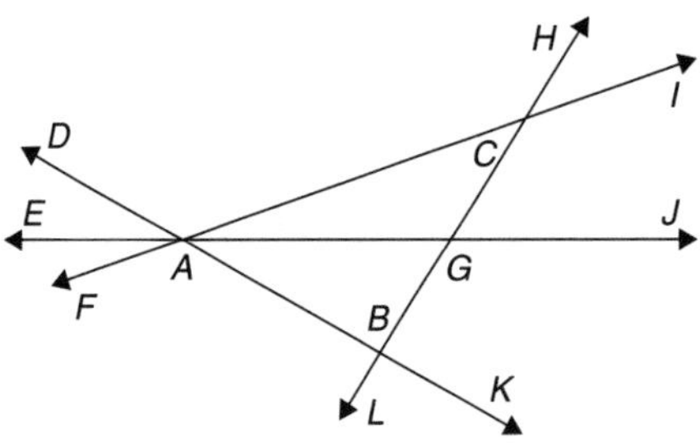

5. Name an angle that forms a pair of vertical angles with ∠HCI.
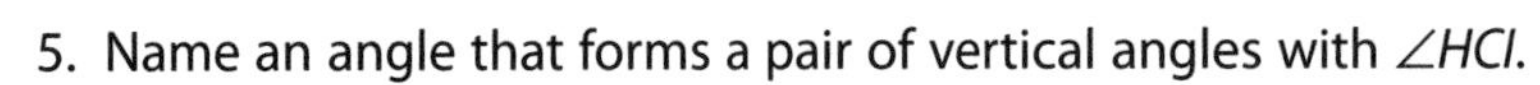
6. Name an angle that forms a linear pair with ∠EAD.
7. Name an angle that forms a pair of vertical angles with ∠GBA.
8. Name an angle that forms a linear pair with ∠GBK.
9. Name an angle that is congruent to ∠BAG.
10. Name an angle that is supplementary to ∠JGC.

Parallel and perpendicular lines

·3·

Lines (and line segments) may be *parallel*, if they are in the same plane and do not intersect, or *skew*, if they are not coplanar and do not intersect. If two lines intersect, they are *coplanar*. If they intersect at right angles, the lines are *perpendicular*.

Parallel lines

Two lines are *parallel* if they are coplanar and never intersect, no matter how far they are extended. To eliminate the possibility of the lines crossing, they must always be the same distance apart. You can't conclude that lines are parallel just because they look parallel. If they were off even a tiny bit, they would eventually intersect if extended far enough. When you need to prove that lines are parallel, you'll use properties other than how they look.

Transversals and angles

A *transversal* is a line that intersects two or more lines at different points. When a transversal intersects any lines, groups of angles are formed. When the lines cut by the transversal are parallel lines, you can make some statements about the relationships among these angles.

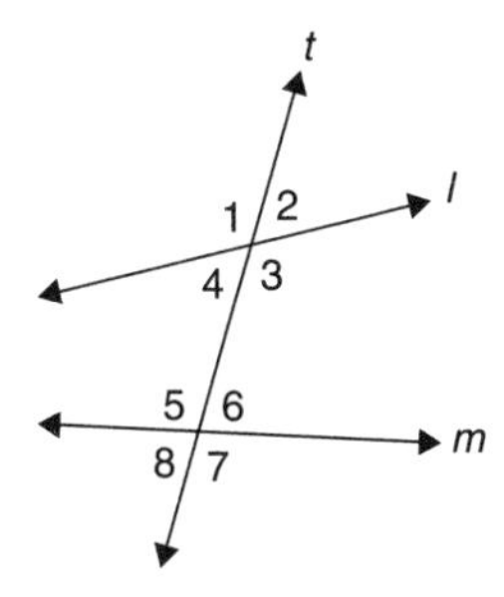

Figure 3.1 Transversal *t* intersects lines *l* and *m*.

As you can see in Figure 3.1, the transversal creates a group of four angles each time it intersects a line. In the figure, line *t* is the transversal. When transversal *t* intersects line *l*, it forms $\angle 1$, $\angle 2$, $\angle 3$, and $\angle 4$. When it intersects line *m*, it forms $\angle 5$, $\angle 6$, $\angle 7$, and $\angle 8$.

Corresponding angles

Two angles, one from each cluster, that are in the same position within their clusters are *corresponding angles*. Examples of corresponding angles are $\angle 1$ and $\angle 5$,

∠2 and ∠6, ∠3 and ∠7, and ∠4 and ∠8. If parallel lines are cut by a transversal, corresponding angles are congruent.

Alternate interior angles

The word *alternate* indicates that the angles are on different sides of the transversal, and the word *interior* means that they are between the two lines. There are two pairs of alternate interior angles in Figure 3.1: ∠4 and ∠6, and ∠3 and ∠5. If parallel lines are cut by a transversal, alternate interior angles are congruent.

Alternate exterior angles

Alternate exterior angles are also positioned on different sides of the transversal, but they are called *exterior* because they are outside the two lines. So ∠1 and ∠7 are alternate exterior angles, as are ∠2 and ∠8. If parallel lines are cut by a transversal, alternate exterior angles are congruent.

Interior angles same side

Examples of interior angles on the same side of the transversal are ∠4 and ∠5 or ∠3 and ∠6 in Figure 3.1. If parallel lines are cut by a transversal, interior angles on the same side of the transversal are supplementary. Note the difference in that statement. The other pairs of angles were congruent, but interior angles on the same side are supplementary.

Use the following diagram to answer these questions.

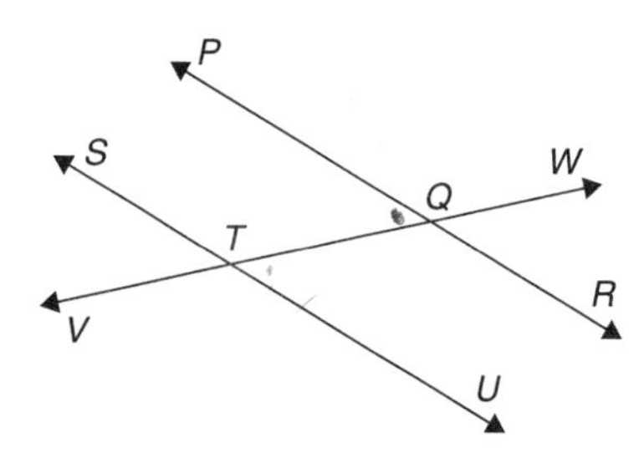

1. ∠*UTQ* and ________ are a pair of corresponding angles.
2. ∠*PQT* and ________ are a pair of alternate interior angles.
3. ∠*VTU* and ________ are a pair of alternate exterior angles.
4. ∠*STQ* and ________ are pair of interior angles on the same side of the transversal.
5. ∠*STQ* and ________ are a pair of vertical angles.
6. ∠*PQT* and ________ are a linear pair.
7. ∠*PQW* and ________ are a pair of corresponding angles.
8. ∠*TQR* and ________ are a pair of alternate interior angles.
9. ∠*STV* and ________ are a pair of alternate exterior angles.
10. ∠*TQR* and ________ are pair of interior angles on the same side of the transversal.

11. If $m\angle STV = 38°$, then $m\angle WQR =$ ________.
12. If $m\angle PQT = 41°$, then $m\angle STQ =$ ________.
13. If $m\angle RQT = 112°$, then $m\angle STQ =$ ________.
14. If $m\angle VTU = 125°$, then $m\angle TQR =$ ________.
15. If $m\angle STQ = 119°$, then $m\angle PQT =$ ________.
16. If $m\angle PQW = 131°$, then $m\angle VTU =$ ________.
17. If $m\angle WQR = 18°$, then $m\angle QTU =$ ________.
18. If $m\angle QTU = 23°$, then $m\angle STV =$ ________.
19. If $m\angle RQT = 107°$, then $m\angle WQR =$ ________.
20. If $m\angle STQ = 99°$, then $m\angle PQW =$ ________.

Parallel and perpendicular lines

If two lines are parallel to the same line, then they are parallel to each other.

If two coplanar lines are perpendicular to the same line, they are parallel to each other.

If a line is parallel to one of two parallel lines, it is parallel to the other.

If a line is perpendicular to one of two parallel lines and lies in the same plane as the parallel lines, it is perpendicular to the other.

EXERCISE
3·2

Fill each blank with the word sometimes, always, *or* never.

1. Line l is parallel to line m. Line m is parallel to line p. Line l is ________ parallel to line p.
2. Line l is parallel to line k. Line k is perpendicular to line q. Line l is ________ parallel to line q.
3. Line a is perpendicular to line b. Line a is perpendicular to line c. Line b is ________ perpendicular to line c.
4. Line d is perpendicular to line f. Line d is parallel to line g. Line f is ________ perpendicular to line g.
5. Line l is perpendicular to line m. Line m is parallel to line p. Line l is ________ perpendicular to line p.
6. Line l is perpendicular to line k. Line k is perpendicular to line q. Line l is ________ parallel to line q.
7. Line a is parallel to line b. Line a is perpendicular to line c. Line b is ________ parallel to line c.
8. Line d is parallel to line f. Line d is parallel to line g. Line f is ________ perpendicular to line g.
9. Line l is parallel to line m. Line m is perpendicular to line p. Line l is ________ perpendicular to line p.
10. Line l is perpendicular to line k. Line k is perpendicular to line q. Line l is ________ parallel to line q.

Proving lines are parallel

To prove that two lines are parallel, you need to prove one of the following is true:

- A pair of corresponding angles is congruent.
- A pair of alternate interior angles is congruent.
- A pair of alternate exterior angles is congruent.
- A pair of interior angles on the same side of the transversal is supplementary.
- Both lines are parallel to the same line.
- Both lines are perpendicular to the same line, and the lines are coplanar.

Determine whether any pair of lines in the following diagram can be proved parallel based on the given information. If so, name the parallel lines and give a reason.

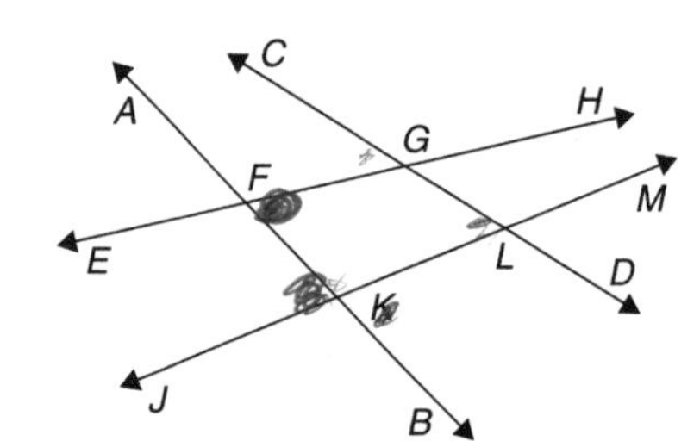

1. $\angle AFE \cong \angle LGH$
2. $\angle AFG \cong \angle GLK$
3. $\angle GLM \cong \angle FKL$
4. $\angle KFG$ and $\angle LGF$ are supplementary.
5. $\angle AFE \cong \angle LKB$
6. $\overleftrightarrow{AB} \perp \overleftrightarrow{JM}$ and $\overleftrightarrow{CD} \perp \overleftrightarrow{JM}$
7. $\angle GLK \cong \angle GFK$
8. $\angle FKL$ and $\angle CGF$ are supplementary.
9. $\angle GLK \cong \angle LKB$
10. $\angle GFK \cong \angle FKJ$

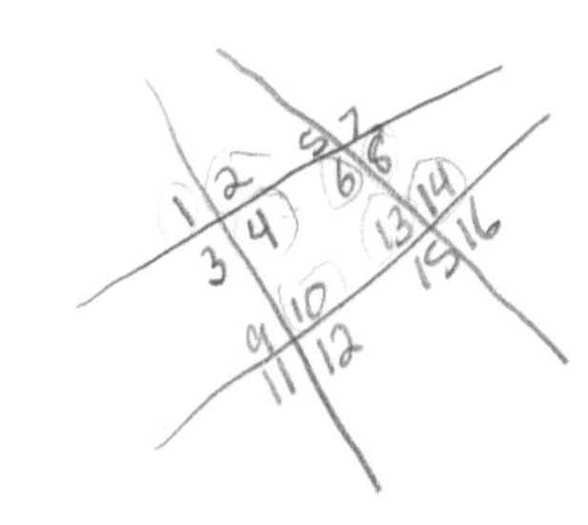

Congruent triangles

•4•

Line segments are congruent if they are the same length, and angles are congruent if they have the same measurement. When you begin to work with more complicated figures, however, you need a new definition of *congruent*. In simple language, however, two figures are congruent if they are exactly the same size and shape.

Congruent polygons

A *polygon* is a closed figure formed from line segments that intersect only at their endpoints. The line segments are the sides of the polygon, and the points where the sides meet are the vertices of the polygon. The polygon has an interior angle at each vertex.

Two polygons are congruent if it is possible to match up their vertices so that corresponding sides are congruent and corresponding angles are congruent. A congruence statement tells which polygons are congruent, for example, $\Delta ABC \cong \Delta XYZ$. It also tells, by the order in which the vertices are named, which sides and angles correspond. The statement $\Delta ABC \cong \Delta XYZ$ indicates that A corresponds to X, B to Y, and C to Z. If $\Delta ABC \cong \Delta XYZ$, $\angle A \cong \angle X$, $\angle B \cong \angle Y$, $\angle C \cong \angle Z$, $\overline{AB} \cong \overline{XY}$, $\overline{BC} \cong \overline{YZ}$, and $\overline{CA} \cong \overline{ZX}$.

EXERCISE

4·1

List the congruent corresponding parts of the congruent polygons.

1. Pentagon $MIXED \cong$ Pentagon $FRUIT$
2. $\Delta CAT \cong \Delta DOG$
3. Quadrilateral $STRA \cong$ Quadrilateral $TREK$. Even though the original $STRA$ loses the Star Trek allusion, T and R must be consecutive vertices.
4. Hexagon $CANDLE \cong$ Hexagon $GARDEN$
5. $\Delta LIP \cong \Delta EAR$

Write a congruence statement based on the given information.

6. $\angle R \cong \angle G$, $\angle S \cong \angle H$, $\angle T \cong \angle I$, $\overline{RS} \cong \overline{GH}$, $\overline{ST} \cong \overline{HI}$, and $\overline{RT} \cong \overline{GI}$

7. $\angle L \cong \angle T$, $\angle LEA \cong \angle TEM$, $\angle A \cong \angle M$, $\angle F \cong \angle S$, $\overline{LE} \cong \overline{TE}$, $\overline{EA} \cong \overline{EM}$, $\overline{AF} \cong \overline{MS}$, and $\overline{LF} \cong \overline{TS}$

8. $\angle E \cong \angle F$, $\angle L \cong \angle C$, $\angle G \cong \angle S$, $\angle O \cong \angle A$, $\angle V \cong \angle R$, $\overline{LO} \cong \overline{CA}$, $\overline{GL} \cong \overline{SC}$, $\overline{GE} \cong \overline{SF}$, $\overline{OV} \cong \overline{AR}$, and $\overline{VE} \cong \overline{RF}$.

9. $\angle R \cong \angle E$, $\angle A \cong \angle L$, $\angle M \cong \angle G$, $\overline{RM} \cong \overline{EG}$, $\overline{AM} \cong \overline{LG}$, and $\overline{AR} \cong \overline{LE}$.

10. $\angle HEM \cong \angle BES$, $\angle O \cong \angle A$, $\angle M \cong \angle S$, $\angle H \cong \angle B$, $\overline{OM} \cong \overline{AS}$, $\overline{ME} \cong \overline{SE}$, $\overline{HO} \cong \overline{BA}$, and $\overline{HE} \cong \overline{BE}$.

Triangles and congruence

Although it is possible to have congruent polygons with any number of sides, most congruence problems involve triangles; so for now, we'll keep the focus there. You'll look at more properties of polygons in a later chapter.

Before we turn to congruent triangles, it's probably a good idea to review some facts about triangles in general. Triangles are classified either by the size of their angles or by the number of congruent sides they have. If the triangle contains a right angle, it's a *right triangle*; and if it contains an obtuse angle, it's an *obtuse triangle*. An *acute triangle* has three acute angles.

A triangle with three congruent sides is an *equilateral triangle*, and one with two congruent sides is *isosceles*. If all sides are different lengths, the triangle is scalene. Equilateral triangles are also *equiangular*; that is, they have three congruent angles, each measuring 60°. If a triangle is isosceles, it has two congruent angles, and those angles sit opposite the congruent sides. In a *scalene triangle*, each angle is a different size.

You probably learned all those facts before you got to a geometry course, but some of them, like the fact that base angles of an isosceles triangle are congruent, are proved by using congruent triangles.

Proving triangles congruent

When triangles are congruent, all the corresponding angles are congruent and all the corresponding sides are congruent. Luckily, it is not necessary to know about every pair of corresponding sides and every pair of corresponding angles to be certain that a pair of triangles is congruent. There are postulates and theorems that tell us the minimum information that will guarantee the triangles are congruent.

- **SSS**: If three sides of one triangle are congruent to the corresponding sides of another triangle, then the triangles are congruent.

  Suppose ΔABC is isosceles with $\overline{AB} \cong \overline{BC}$ and M is the midpoint of $\overline{AC}$. If you draw the median from B to M, you have

  - $\overline{AB} \cong \overline{BC}$ because the triangle is isosceles (S).
  - $\overline{AM} \cong \overline{MC}$ because the midpoint divides $\overline{AC}$ into two congruent segments (S).
  - $\overline{BM} \cong \overline{BM}$ by the reflexive property (S).

  So $\Delta ABM \cong \Delta CBM$ by SSS.

- **SAS**: If two sides and the angle included between them in one triangle are congruent to the corresponding parts of another triangle, then the triangles are congruent.

  Start again with isosceles ΔABC with $\overline{AB} \cong \overline{BC}$, and draw the bisector of $\angle B$. Call it $\overline{BD}$. This creates two triangles.

- $\overline{AB} \cong \overline{BC}$ because the original triangle is isosceles (S).
- $\angle ABD \cong \angle CBD$ because the angle bisector divides $\angle B$ into two congruent angles (A).
- $\overline{BD} \cong \overline{BD}$ by the reflexive property (S).

So $\Delta ABD \cong \Delta CBD$ by SAS.

- **ASA**: If two angles and the side included between them in one triangle are congruent to the corresponding parts of another triangle, then the triangles are congruent.

Suppose that, in ΔRST, there is a line segment $\overline{SP}$ that bisects $\angle S$ and $\overline{SP} \perp \overline{RT}$. (If you think that's a lot to ask one segment to do, you're right, and you should never assume that a segment does more than one job. In this case, you're given that fact, so it's OK.)

- $\overline{SP}$ bisects $\angle S$, $\angle RSP \cong \angle TSP$ (A).
- $\overline{SP} \cong \overline{SP}$ by the reflexive property (S).
- So you know $\overline{SP} \perp \overline{RT}$ and that means $\angle SPR$ and $\angle SPT$ are right angles, and $\angle SPR \cong \angle SPT$ because all right angles are congruent (A).

So $\Delta RSP \cong \Delta TSP$ by ASA.

- **AAS**: If two angles and a side not included between them in one triangle are congruent to the corresponding parts of another triangle, then the triangles are congruent.

AAS is actually a theorem, rather than a postulate. You prove it's true by showing that whenever you have two angles of one triangle congruent to two angles of another, the third angles are congruent as well, because the three angles of a triangle always add to 180°. If you have AAS, you can quickly show that ASA is true as well, so we accept AAS.

- **HL**: If the hypotenuse and a leg of one right triangle are congruent to the hypotenuse and corresponding leg of another right triangle, then the right triangles are congruent.

If ΔABC is isosceles with $\overline{AB} \cong \overline{BC}$ and you draw $\overline{BE} \perp \overline{AC}$, you form two triangles: ΔABE and ΔEBC.

- Since $\overline{BE} \perp \overline{AC}$, both triangles are right triangles.
- $\overline{AB}$ and $\overline{BC}$ are the hypotenuses of the right triangles and $\overline{AB} \cong \overline{BC}$ (H).
- Since $\overline{BE}$ is a side of both triangles and congruent to itself by the reflexive property, $\overline{BE} \cong \overline{BE}$ (L).

You have the hypotenuse and leg of one right triangle congruent to the corresponding parts of the other triangle, so the two right triangles are congruent by HL. State the congruence to show what matches: $\Delta ABE \cong \Delta CBE$.

EXERCISE

4·2

Decide if the information given is enough to guarantee that the triangles are congruent. If so, write a congruence statement and state the postulate or theorem that guarantees the congruence. If the given information is not adequate, write "Cannot be determined."

1. $\angle A \cong \angle X$, $\angle B \cong \angle Y$, $\angle C \cong \angle Z$, $\overline{AB} \cong \overline{XY}$
2. $\overline{CA} \cong \overline{DO}$, $\overline{AT} \cong \overline{OG}$, $\overline{CT} \cong \overline{DG}$
3. $\overline{BO} \cong \overline{CA}$, $\angle O \cong \angle A$, $\overline{OX} \cong \overline{AR}$
4. $\overline{BI} \cong \overline{MA}$, $\overline{IG} \cong \overline{AN}$, $\angle B \cong \angle M$
5. $\overline{CT} \cong \overline{IN}$, $\angle A \cong \angle W$, $\angle C \cong \angle I$
6. $\overline{AG} \cong \overline{CE}$, $\overline{BA} \cong \overline{IC}$, $\angle A \cong \angle C$
7. $\angle G \cong \angle R$, $\angle B \cong \angle J$, $\angle U \cong \angle A$
8. $\angle A \cong \angle P$, $\angle R \cong \angle E$, $\overline{RT} \cong \overline{EN}$
9. $\angle O \cong \angle A$, $\overline{GT} \cong \overline{LP}$, $\overline{OT} \cong \overline{AP}$
10. $\overline{OY} \cong \overline{AM}$, $\overline{TY} \cong \overline{JM}$, $\overline{TO} \cong \overline{JA}$

CPCTC

If two triangles (or other polygons) are congruent, all corresponding sides and all corresponding angles are congruent. Once you've used one of the minimum requirements in the rules above to show that two triangles are congruent, the remaining angles and sides are also congruent to their corresponding partners. This is usually stated as *corresponding parts of congruent triangles are congruent,* abbreviated as CPCTC.

Remember when we took isosceles ΔABC and drew the median from B to M, the midpoint of $\overline{AC}$, and proved $\Delta ABM \cong \Delta CBM$ by SSS? That involved showing that $\overline{AB} \cong \overline{BC}$, $\overline{AM} \cong \overline{MC}$, and $\overline{BM} \cong \overline{BM}$. Now that you know the triangles are congruent, CPCTC tells you that $\angle A \cong \angle C$, $\angle ABM \cong \angle CBM$, and $\angle BMA \cong \angle BMC$. That proves that the base angles of an isosceles triangle—in this case, $\angle A$ and $\angle C$, the angles opposite the congruent sides—are congruent. Since $\angle ABM \cong \angle CBM$, the median $\overline{AM}$ is also the bisector of $\angle B$. With a little work, you can also show that because $\angle BMA \cong \angle BMC$ and they form a linear pair, they must both be right angles, and so $\overline{AM} \perp \overline{AC}$, which makes it an altitude of the triangle.

- If two sides of a triangle are congruent, the angles opposite those sides are congruent. (You can prove that the converse of this is also true: If two angles of a triangle are congruent, the sides opposite those angles are congruent. Try drawing the bisector of $\angle B$ and proving the triangles congruent by AAS.)
- In an isosceles triangle, the median to the base, the altitude to the base, and the bisector of the vertex angle are all the same line segment.

The information given is sufficient to prove a pair of triangles congruent. Write the congruence statement and the postulate or theorem that guarantees congruence. Then list the remaining pairs of congruent corresponding parts.

1. $\overline{EA} \cong \overline{KI}$, $\overline{AR} \cong \overline{IN}$, $\overline{ER} \cong \overline{KN}$
2. $\overline{LI} \cong \overline{FE}$, $\overline{IP} \cong \overline{EW}$, $\angle I \cong \angle E$
3. $\angle E \cong \angle O$, $\angle G \cong \angle R$, $\overline{EG} \cong \overline{OR}$, $\overline{LG} \cong \overline{FR}$
4. $\overline{AM} \cong \overline{LW}$, $\angle R \cong \angle O$, $\angle M \cong \angle W$
5. $\overline{TE} \cong \overline{FR}$, $\overline{OE} \cong \overline{AR}$, $\angle E \cong \angle R$

Determine whether the given information is sufficient to prove the triangles congruent. If so, write the congruence statement and the postulate or theorem that guarantees congruence. Then list the remaining pairs of congruent corresponding parts. If the given information is not adequate, write "Cannot be determined."

6. $\overline{BY} \cong \overline{MN}$, $\overline{BO} \cong \overline{ME}$, $\overline{OY} \cong \overline{EN}$
7. $\overline{MO} \cong \overline{PA}$, $\overline{OW} \cong \overline{AN}$, $\angle M \cong \angle P$
8. $\overline{CN} \cong \overline{SW}$, $\angle NCA \cong \angle WSE$, $\angle ANC \cong \angle EWS$
9. $\overline{DM} \cong \overline{TP}$, $\overline{DI} \cong \overline{TA}$, $\angle D \cong \angle T$
10. $\overline{GE} \cong \overline{WH}$, $\overline{ET} \cong \overline{HO}$, $\overline{GT} \cong \overline{WO}$

Two-stage proofs

There are times when you are asked to prove a pair of triangles congruent, but you find that the given information is not sufficient. In those cases, it is often possible to first prove a different pair of triangles congruent and then use CPCTC to obtain the information you need to prove that the desired triangles are congruent.

Name the pair of triangles that can be proved congruent immediately and the postulate or theorem that guarantees that congruence.

1.

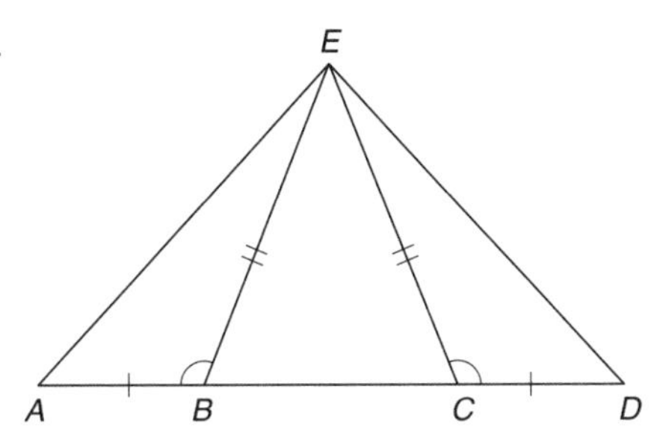

2.

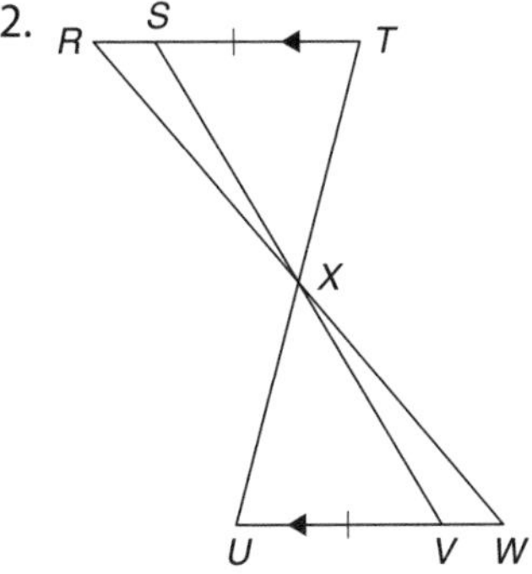

3.

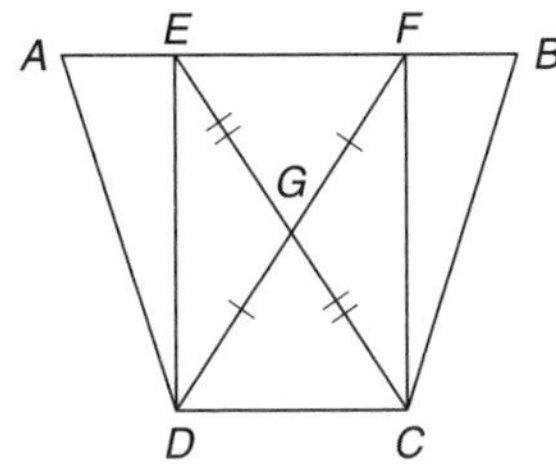

Determine (a) which triangles can easily be proved congruent and (b) which congruent corresponding parts would be necessary to prove the specified triangles congruent.

4. Prove $\Delta CAE \cong \Delta ACF$.

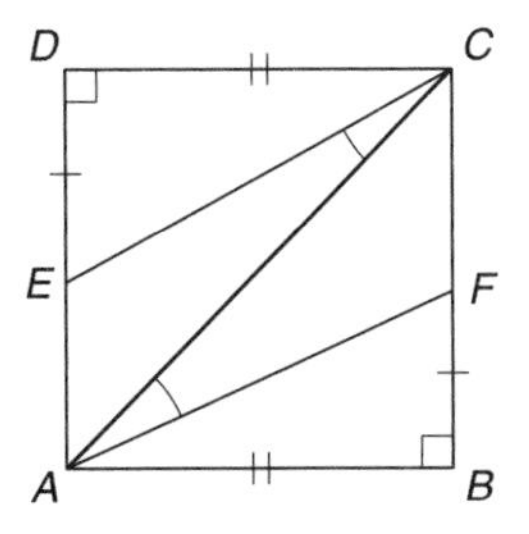

5. Prove $\triangle BDE \cong \triangle RED$.

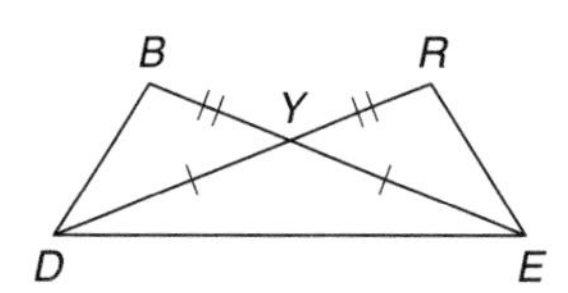

6. Given $AB = CD$, $HB = DF$. Prove $\triangle HAB \cong \triangle FCD$.

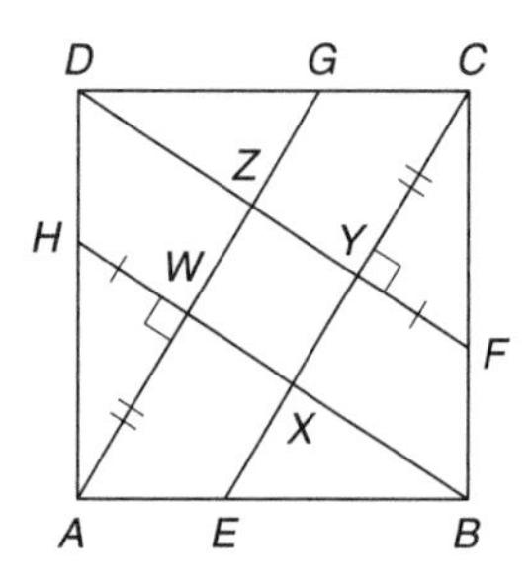

7. Prove $\triangle BAD \cong \triangle ABC$.

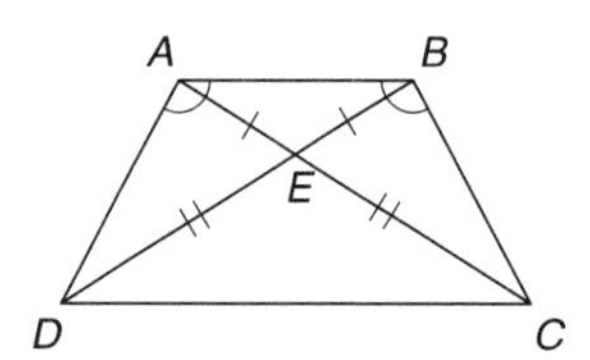

8. Given $\overline{DC} \parallel \overline{AB}$. Prove $\triangle ADE \cong \triangle BCF$.

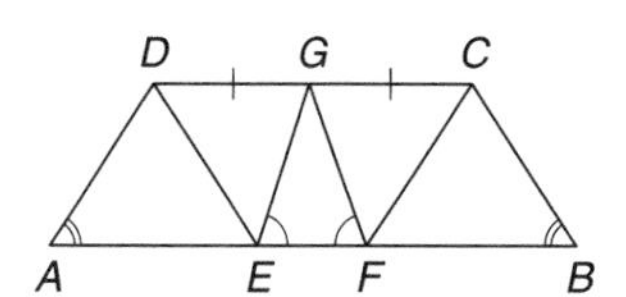

9. Prove $\triangle PQR \cong \triangle PTS$.

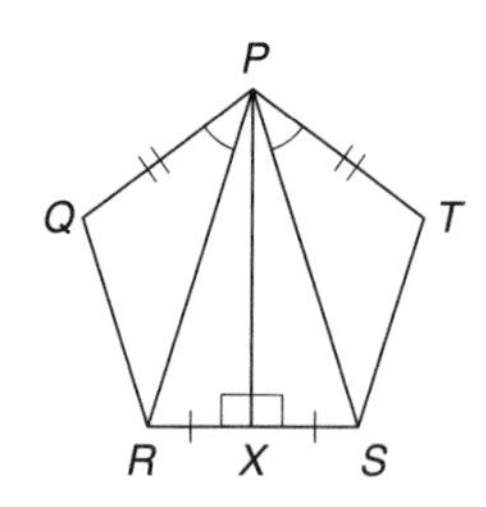

10. Given C is the midpoint of $\overline{AE}$. Prove $\triangle ABC \cong \triangle CDE$.

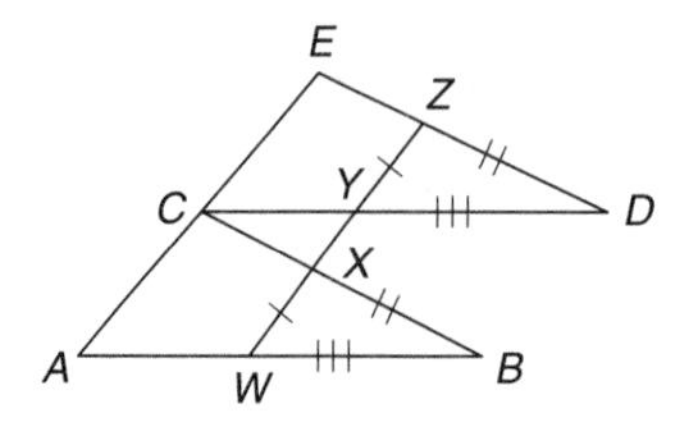

Congruent triangle proofs

You can organize a proof in two columns with statements in one column and supporting reasons in the other, or you can simply set your ideas out in paragraph form. Different teachers will specify different rules about the amount of detail they expect and whether you can refer to theorems by names or you should write out the entire statement of the theorem. Whatever format you choose and whatever rules you follow, it's a good idea to map out a plan for the major steps in the proof before you begin to write.

Make a plan for each proof.

1. Given ΔBEC is isosceles with $\overline{BE} \cong \overline{CE}$ and $\overline{AB} \cong \overline{CD}$. Prove $\Delta CAE \cong \Delta BDE$.

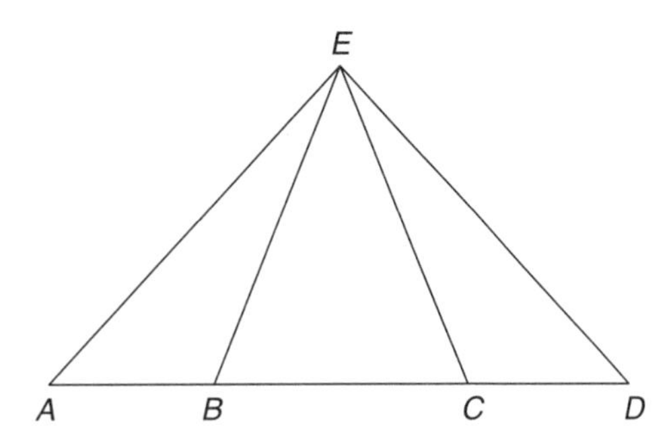

2. Given $\overline{SV}$ and $\overline{RW}$ bisect each other. Prove $\angle T \cong \angle U$.

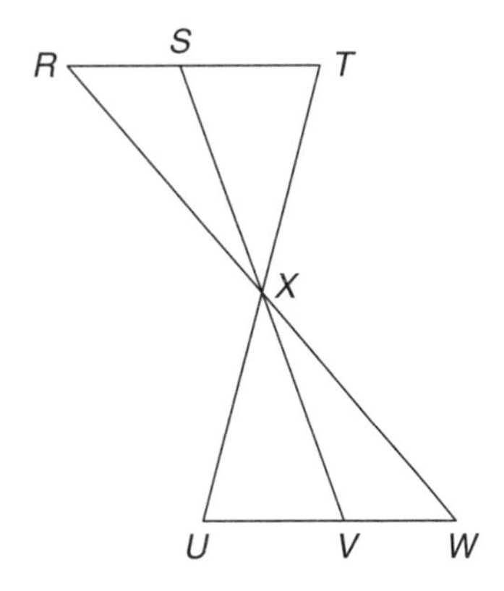

3. Given $\overline{DE} \perp \overline{AB}$, $\overline{FC} \perp \overline{AB}$, $\overline{ED} \cong \overline{FC}$, and $\overline{AE} \cong \overline{FB}$. Prove $\Delta ADF \cong \Delta BCE$.

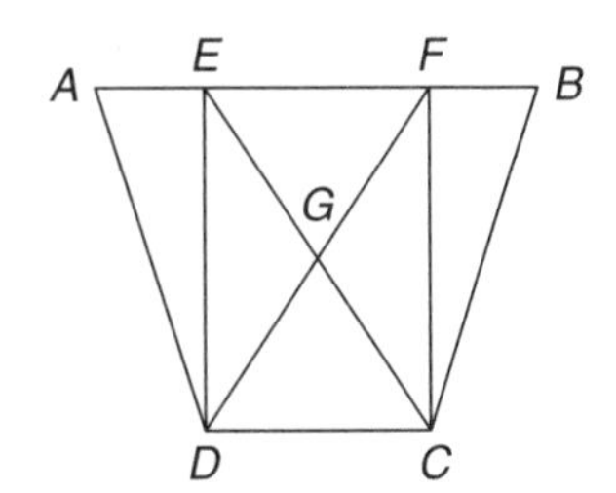

4. Given square $ABCD$ with $\overline{CD} \perp \overline{DE}$, $\overline{AB} \perp \overline{BF}$, $\overline{EA} \cong \overline{CF}$, and $\overline{DE} \cong \overline{FB}$. Prove $\angle FAC \cong \angle ECA$.

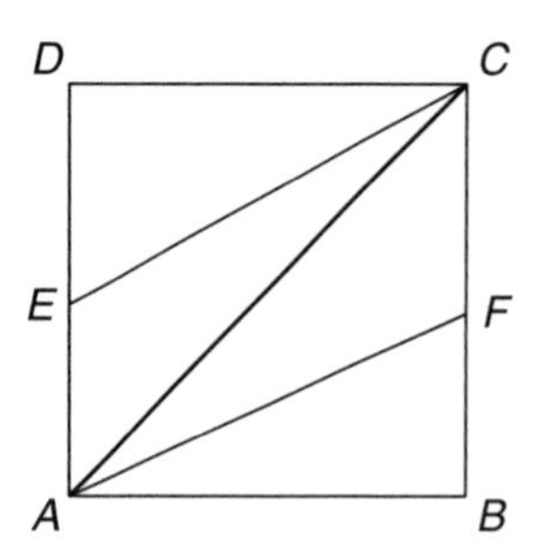

5. Given $\overline{BD} \cong \overline{RE}$, $\overline{BY} \cong \overline{RY}$, and $\angle B \cong \angle R$. Prove $\Delta DEB \cong \Delta EDR$.

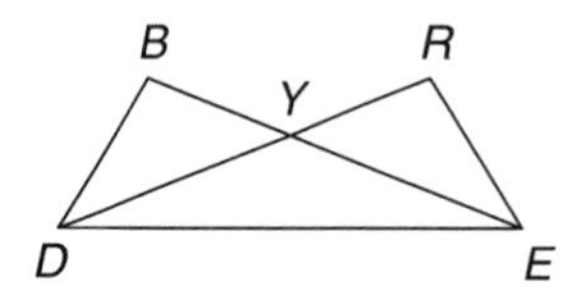

Inequalities

·5·

Congruence deals with figures that are exactly the same size and shape, but in geometry, there are also important relationships that talk about sides and angles that are unequal.

Inequalities in one triangle

The first group of inequalities deals with the sides of a single triangle and the relationship between the sides and angles in a single triangle.

The triangle inequality

In any triangle, the sum of the lengths of two sides is greater than the length of the third side. The shortest distance between two points is along a line, so moving around the triangle from vertex A to B and then from B to C will be longer than AC.

If a, b, and c are the sides of a triangle, the fact that $a+b>c$ also means that $a>|c-b|$. The triangle inequality means that the length of any side of a triangle is less than the sum of the other two sides and greater than their difference.

Tell whether it is possible for the given lengths to be the sides of a triangle.

1. 5 cm, 8 cm, 9 cm
2. 3 in, 7 in, 8 in
3. 12 m, 18 m, 31 m
4. 15 ft, 23 ft, 23 ft
5. 5 mm, 5 mm, 10 mm

Given the lengths of two sides, find the range of values for the third side of the triangle.

6. $a=5$ cm, $b=8$ cm, ________ $<c<$ ________
7. $p=4$ cm, $q=11$ cm, ________ $<r<$ ________

8. $x = 12$ cm, $y = 21$ cm, ________ $< z <$ ________

9. $d = 24$ cm, $e = 19$ cm, ________ $< f <$ ________

10. $r = 7$ cm, $s = 18$ cm, ________ $< t <$ ________

Sides and angles

In a triangle, the longest side lies opposite the largest angle. In a right triangle, for example, the hypotenuse is the longest side of the triangle and lies opposite the 90° angle. The smallest angle of any triangle is opposite the shortest side. If two sides of a triangle are congruent, the angles opposite those sides are congruent; and conversely, if a triangle is scalene, with sides $AB = 4$ cm, $BC = 7$ cm, and $AC = 9$ cm, then $\angle C$ is the smallest angle because it's opposite the shortest side. Here $\angle B$ is the largest angle, because it's opposite the longest side, and $\angle A$ is somewhere in between.

Name the sides of the triangle in order from shortest to longest.

1. In ΔABC, $m\angle A = 42°$, $m\angle B = 18°$
2. In ΔRST, $m\angle R = 49°$, $m\angle S = 35°$
3. In ΔXYZ, $m\angle X = 26°$, $m\angle Y = 68°$
4. In ΔDEF, $m\angle D = 28°$, $m\angle F = 43°$
5. In ΔPQR, $m\angle Q = 68°$, $m\angle R = 21°$

Name the angles of the triangle in order from largest to smallest.

6. $PQ = 23$ cm, $QR = 12$ cm, $PR = 13$ cm
7. $AB = 8$ cm, $BC = 13$ cm, $AC = 10$ cm
8. $XY = 3$ cm, $YZ = 4$ cm, $XZ = 6$ cm
9. $RS = 12$ cm, $ST = 13$ cm, $RT = 5$ cm
10. $DE = 6$ cm, $EF = 11$ cm, $DF = 16$ cm

Putting sides in order

When a figure is composed of two or more triangles that share sides, you may be able to use the information about how the lengths of sides and the sizes of angles in each triangle are related to put the sides of the whole figure in order. Focus on one triangle and put the sides of that triangle in order, and then move to the other triangle and do the same. If the shared side is the largest side of one triangle and the smallest side of the other, you can link the two inequalities and order all five segments.

Put the segments in each figure in order from shortest to longest.

1. 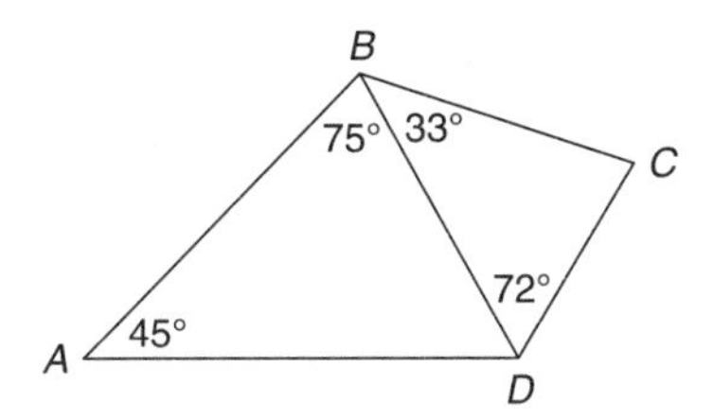

2. 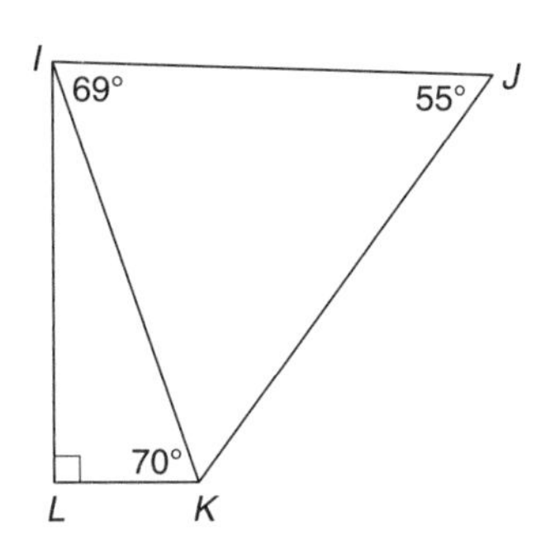

3. 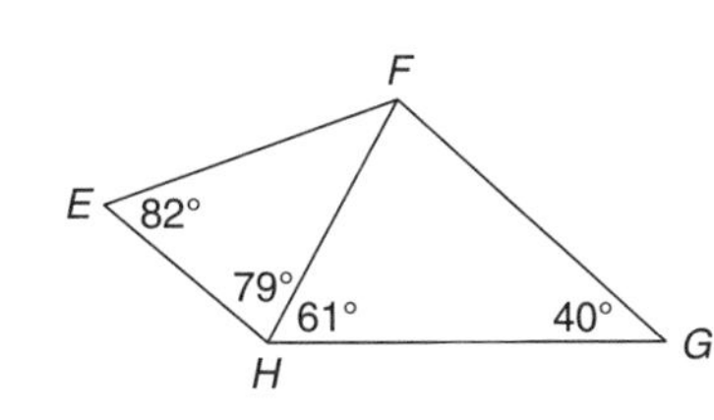

4. 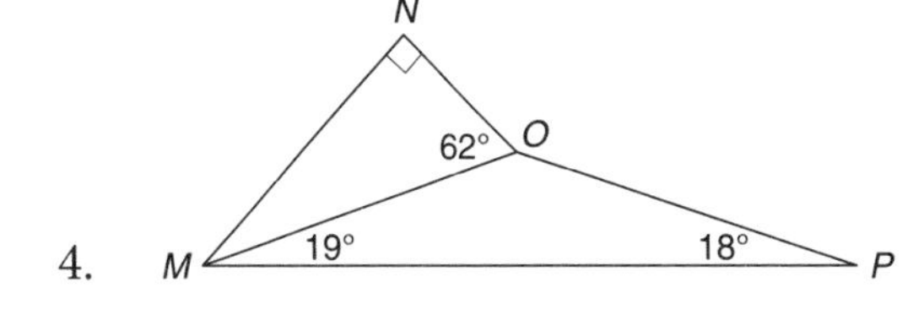

5. 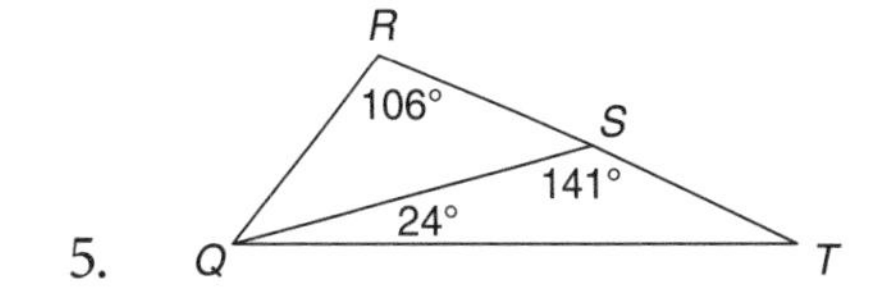

Inequalities in two triangles

When you study congruent triangles, you are interested in proving that two triangles are identical, and you often use the side-angle-side, or SAS, theorem to prove that two triangles are congruent. At other times, however, the differences between triangles are important.

Hinge theorem

If two sides of one triangle are congruent to the corresponding sides of another triangle, but the included angles are *not* congruent, then the triangle with the larger included angle has the longer third side.

Fill in each blank with $>$, $<$, or $=$.

1. $AB = DE = 5$ cm, $BC = EF = 8$ cm, $m\angle B = 59°$, $m\angle E = 61°$. AC ________ DF
2. $RT = MP = 2$ ft, $ST = PN = 3$ ft, $m\angle T = 102°$, $m\angle P = 98°$. RS ________ MN
3. $XY = RS = 12$ in, $XZ = ST = 15$ in, $m\angle X = 84°$, $m\angle S = 87°$. YZ ________ RT
4. $PQ = AC = 4$ m, $PR = CB = 7$ m, $m\angle P = 110°$, $m\angle C = 108°$. QR ________ AB

5. $ML = 8$ in, $MN = 15$ in, $m\angle M = 50°$, $PO = 15$ in, $m\angle P = 40°$, $QO = 8$ in, $\overline{OQ} \perp \overline{QP}$.
   LN ________ QP

6. $m\angle B = 43°$, $BC = AC = 9.5$ cm, $AB = 13$ cm, $FE = 9.5$ cm, $m\angle E = 94°$, $FD = 13$ cm, $m\angle F = 48°$.
   CA ________ ED

7. $m\angle X = 25°$, $m\angle Z = 40°$, $m\angle S = 40°$, $m\angle T = 115°$, $XZ = 15$, $YZ = 5$, $RS = 15$, $TS = 5$.
   XY ________ RT

8. ΔFGH is isosceles with $FG = FH$. $\overline{FK}$ bisects $\angle F$. GK ________ KH

9. ΔRST is a right triangle with $\overline{TS} \perp \overline{SR}$. V is the midpoint of $\overline{TR}$ and $m\angle SVT > m\angle SVR$.
   ST ________ SR

10. ΔMNP has an acute angle at P. Side $\overline{MP}$ is extended through P to O so that $MP = PO$, and $\overline{NO}$ is drawn. MN ________ NO

Converse of hinge theorem

Suppose two sides of one triangle are congruent to the corresponding sides of another triangle, but the third side of the first is larger than the third side of the other. The angle included between the congruent corresponding sides is larger in the triangle with the larger third side.

Fill in each blank with >, <, or =.

1. $AB = DE = 5$ cm, $BC = EF = 8$ cm, $AC = 10$ cm, $DF = 11$ cm. $m\angle B$ ______ $m\angle E$
2. $RS = XY = 8$ in, $ST = YZ = 15$ in, $RT = 17$ in, $XZ = 19$ in. $m\angle S$ ________ $m\angle Y$
3. $XY = PQ$, $XZ = PR$, $YZ > QR$. $m\angle X$ ______ $m\angle P$
4. $WV = MN$, $UV = NO$, $WU < MO$. $m\angle V$ ______ $m\angle N$
5. $AB = DE$, $BC > EF$, $AC = DF$. $m\angle A$ ______ $m\angle D$
6. In equilateral triangle ΔRST, V is the midpoint of $\overline{ST}$. $m\angle RVS$ ________ $m\angle RVT$
7. ΔADB is isosceles with $DA = DB$. Side $\overline{AB}$ is extended through B to C and $\overline{DC}$ is drawn.
   $m\angle ADC$ ________ $m\angle ADB$
8. ΔRST is an isosceles triangle with $\overline{RS} \cong \overline{ST}$. Point V lies on RT and $RV < VT$. $m\angle RSV$
   ________ $m\angle VST$
9. ΔABC is a right triangle with $\overline{AB} \perp \overline{BC}$. $\overline{BD}$ is drawn to the midpoint of $\overline{AC}$. $m\angle C = 48°$.
   $m\angle BDC$ ________ $m\angle BDA$
10. In ΔJKL, $\overline{KM}$ is drawn to the midpoint of $\overline{JL}$. $JK < KL$. $m\angle KMJ$ ________ $m\angle KML$

Quadrilaterals and other polygons

A *polygon* is a closed figure formed from line segments that intersect only at their endpoints. The line segments are the sides of the polygon, and the points where the sides meet are the vertices of the polygon. The polygon has an interior angle at each vertex.

Polygons are given specific names according to the number of their sides. Common polygons include these:

3 sides: triangle
4 sides: quadrilateral
5 sides: pentagon
6 sides: hexagon
8 sides: octagon
10 sides: decagon

If all the sides of a polygon are congruent, the polygon is equilateral. If all the interior angles of the polygon have the same measure, the polygon is equiangular. A polygon that is both equilateral and equiangular is a regular polygon. When a triangle is equilateral, it is automatically equiangular, but that is not the case with other polygons, so don't assume it.

Diagonals and convex polygons

A diagonal is a line segment connecting two nonadjacent vertices of a polygon. A triangle has no diagonals. Quadrilaterals have two diagonals, and pentagons have five. If the polygon has n sides, it has $n(n-3)/2$ diagonals.

A polygon is convex if all its diagonals fall within the polygon. If a polygon is not convex, it is concave. Concave polygons "cave in," or seem to be dented.

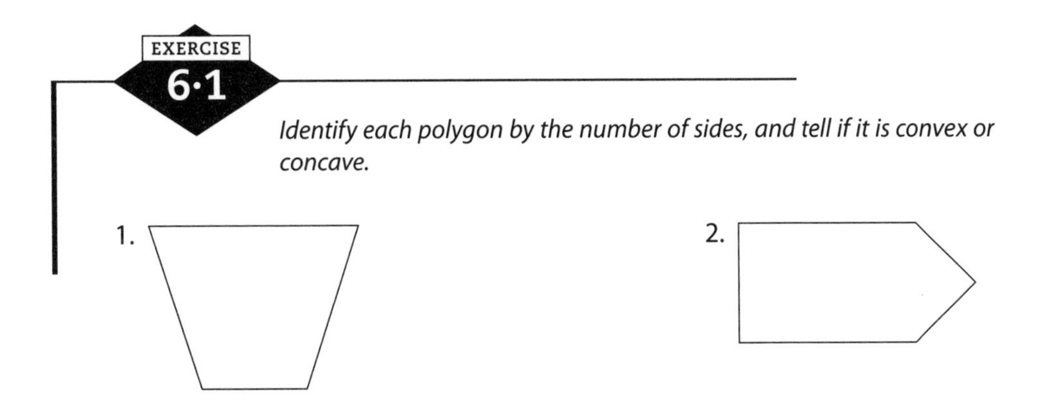

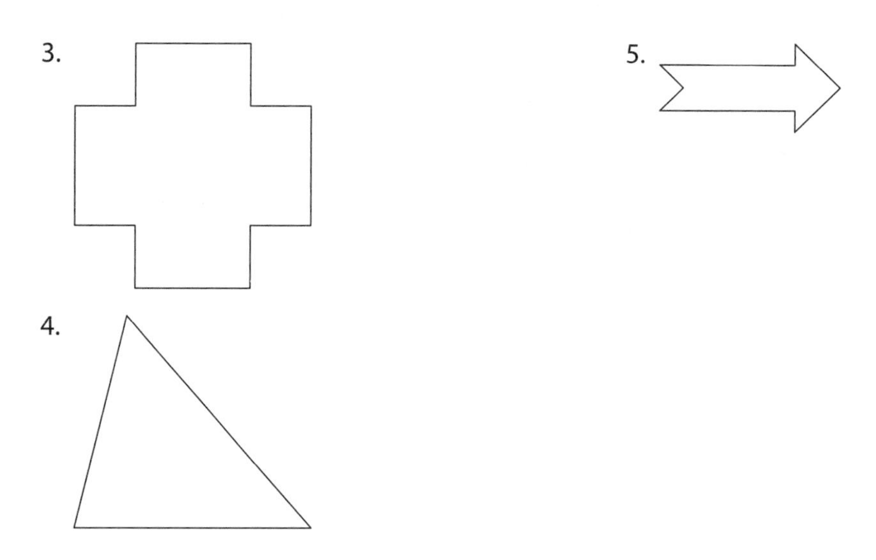

Find the number of diagonals in each polygon.

6. Hexagon
7. Octagon
8. Decagon
9. 11-gon
10. 19-gon

Angles in polygons

The total number of degrees in all the angles of a polygon varies with the number of sides.

Interior angles

The three angles of a triangle add to 180°. A convex polygon with more than three sides can be divided into triangles by drawing all possible diagonals from one vertex. The measures of the angles in all those triangles add up to the measures of the interior angles of the polygon. A quadrilateral divides into two triangles; its interior angles add to $2(180°) = 360°$. In general, the number of triangles created is 2 less than the number of sides, so the sum of the interior angles of a polygon with n sides is $(n - 2) \cdot 180°$. It takes more work, but this can also be shown to be true for concave polygons.

If the polygon is regular and all the interior angles are equal size, you can divide the total by the number of angles to find the measure of one interior angle. In a regular polygon with n sides, the measure of one interior angle is $(n-2) \cdot 180°/n$.

Exterior angles

An exterior angle of a polygon is formed by extending one side of the polygon through one vertex and beyond. The exterior angle formed in this way is supplementary to the interior angle at that vertex, because the angles form a linear pair.

In a triangle, an exterior angle and the adjacent interior angle add to 180°, and the three angles of a triangle add to 180°. As a result, you can show that the measure of an exterior angle of a triangle is equal to the sum of the two remote interior angles. Since the exterior angle is equal to the two remote interior angles combined, it's larger than either one of them.

In any polygon, if one exterior angle is created at each vertex and the measures of these exterior angles are added, the total is always 360°, no matter how many sides the polygon has. If the polygon is regular, all the interior angles are congruent, and therefore all the exterior angles are congruent. In a regular polygon of n sides, each exterior angle measures $360/n$ degrees.

Find the total number of degrees in the interior angles of the polygon described.

1. Pentagon
2. Octagon
3. Hexagon
4. Decagon
5. 18-gon

Find the measure of one interior angle of a regular polygon with n *sides.*

6. $n = 8$
7. $n = 12$
8. $n = 6$
9. $n = 20$
10. $n = 5$

Find the specified measurement.

11. An exterior angle of a regular hexagon
12. The exterior angle at the fifth vertex of a pentagon, if the exterior angles at the other four vertices measure 70°, 74°, 82°, and 61°
13. An exterior angle of a regular polygon with 15 sides
14. The exterior angle at the 12th vertex of a dodecagon if the other 11 exterior angles total 239°
15. The number of sides in a regular polygon if the measure of each exterior angle is 40°

Quadrilaterals

A *quadrilateral* is a polygon with four sides. The four angles of a quadrilateral have measurements that total 360°.

EXERCISE 6·3

Find the missing angle measurements.

1. In quadrilateral *ABCD*, $m\angle A = 84°$, $m\angle B = 117°$, and $m\angle C = 63°$. Find $m\angle D$.
2. In quadrilateral *DEFG*, $m\angle D = 100°$, $m\angle F = 100°$, and $m\angle G = 72°$. Find $m\angle E$.
3. In quadrilateral *WXYZ*, $m\angle W = 73°$, $m\angle X = m\angle Y$, and $m\angle Z = 107°$. Find $m\angle X$.
4. In quadrilateral *PQRS*, $m\angle P = 158°$, $m\angle Q = (4x - 3)°$, $m\angle R = (6x + 3)°$, and $m\angle S = (3x - 6)°$. Find $m\angle R$.
5. In quadrilateral *JKLM*, $m\angle J = x°$, $m\angle K = 2x°$, $m\angle L = 3x°$, and $m\angle M = 4x°$. Find $m\angle J$.

Trapezoids

A *trapezoid* is a quadrilateral with one pair of parallel sides. The parallel sides are called the *bases.* Angles formed by the parallel sides and one nonparallel side are consecutive angles, and consecutive angles are supplementary.

If the nonparallel sides are congruent, the trapezoid is an isosceles trapezoid. The angles at each end of one of the parallel sides are called *base angles.* Base angles of an isosceles trapezoid are congruent. The diagonals of an isosceles trapezoid are congruent.

The line segment that connects the midpoints of the nonparallel sides of a trapezoid is called the *median* of the trapezoid. (Sometimes it's called the *midsegment.*) The median is parallel to the bases. The length of the median is the average of the lengths of the parallel bases.

ABCD *is a trapezoid with* $\overline{AB} \parallel \overline{DC}$. *Find the measurement of the indicated angle.*

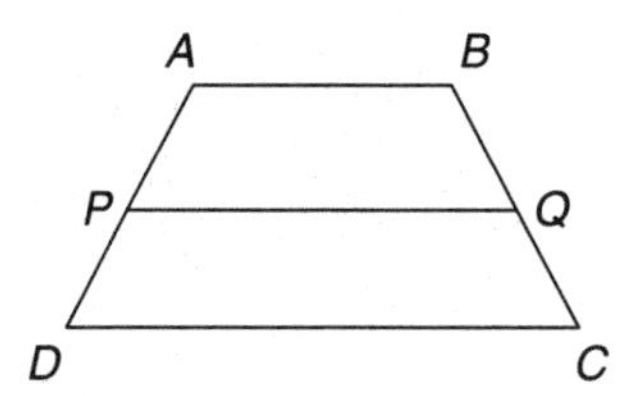

1. $m\angle A = 84°$ and $m\angle C = 63°$. Find $m\angle D$.
2. $m\angle C = 3x - 1°$, $m\angle B = 12x + 1°$, $m\angle D = 7x - 4°$, and $m\angle A = 8x + 4°$. Find $m\angle B$.
3. $\overline{AD} \cong \overline{BC}$ and $m\angle D = 73°$. Find $m\angle C$.
4. $\overline{AD} \cong \overline{BC}$ and $m\angle C = 65°$. Find $m\angle A$.
5. $\overline{AD} \cong \overline{BC}$, $m\angle B = 7x - 5°$, and $m\angle D = 5x - 7°$. Find $m\angle B$.

ABCD *is a trapezoid with* $\overline{AB} \parallel \overline{DC}$ *and median* $\overline{PQ}$. *Find the length of the indicated segment.*

6. $AB = 12$ cm and $CD = 28$ cm. Find PQ.
7. $AB = 8$ in and $PQ = 15$ in. Find CD.
8. $AB = x + 2$ cm, $CD = 29$ cm, and $PQ = 2x - 7$ cm. Find PQ.
9. $AB = 2x + 5$ in, $CD = 5x - 3$ in, and $PQ = 4x - 5$ in. Find PQ.
10. $AB = 8x + 1$ mm, $CD = 3x + 14$ mm, and $PQ = 6x - 10$ mm. Find AB.

Parallelograms

A *parallelogram* is a quadrilateral with two pairs of parallel opposite sides. Opposite sides of a parallelogram are congruent, and opposite angles are congruent as well. Consecutive angles of a parallelogram are supplementary, and the diagonals of a parallelogram bisect each other.

A quadrilateral is a parallelogram if

- both pairs of opposite sides are parallel
- both pairs of opposite sides are congruent
- one pair of opposite sides is both parallel and congruent
- both pairs of opposite angles are congruent
- the diagonals of the quadrilateral bisect each other

EXERCISE **6·5**

Tell whether the quadrilateral is a parallelogram.

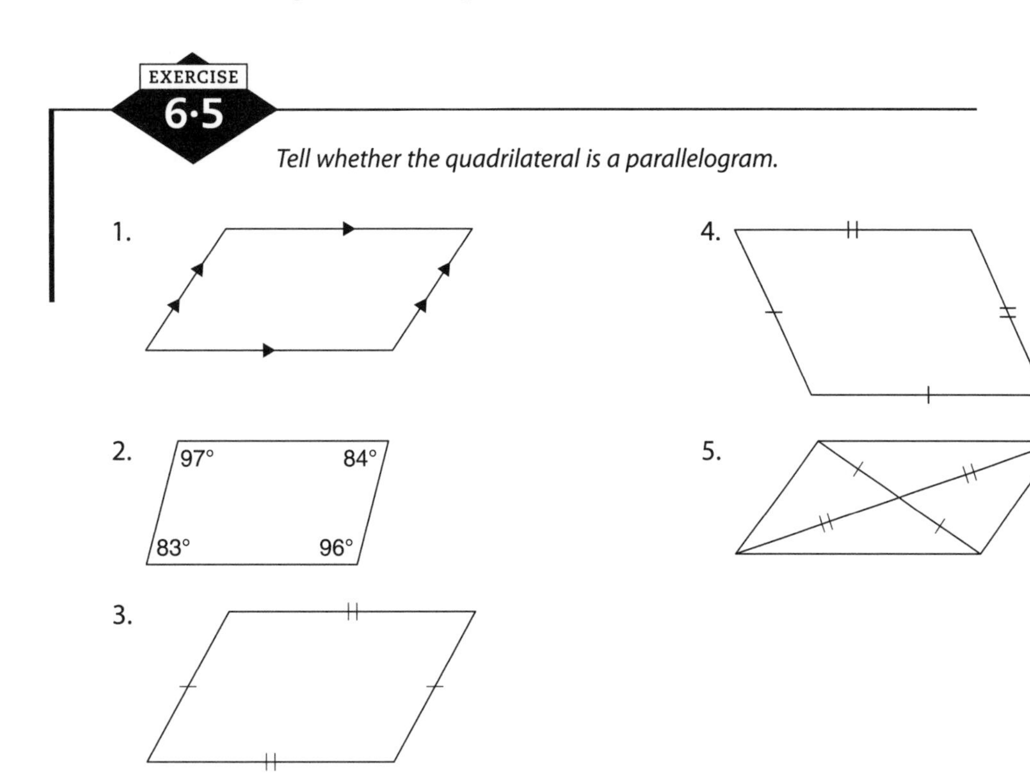

ABCD *is a parallelogram. Find the measurement of the indicated angles.*

6. $m\angle A = 52°$. Find $m\angle B$, $m\angle C$, and $m\angle D$.
7. $m\angle A = 93°$. Find $m\angle B$, $m\angle C$, and $m\angle D$.
8. $m\angle A = 12x - 1°$ and $m\angle B = 8x + 1°$. Find $m\angle C$ and $m\angle D$.
9. $m\angle A = 8x - 9°$ and $m\angle C = 6x + 21°$. Find $m\angle B$ and $m\angle D$.
10. $m\angle A = x^2 - 12°$ and $m\angle C = 9x + 10°$. Find $m\angle B$ and $m\angle D$.

ABCD *is a parallelogram. Find the length of the indicated segment.*

11. $AB = 12$ cm and $BC = 8$ cm. Find CD.
12. $AB = 4$ in and $BC = 9$ in. Find AD.
13. $AD = 6x - 1$ cm, $CD = 4x - 1$ cm, and $BC = 5x + 3$ cm. Find AB.
14. $AB = 3(x - 1)$ cm, $CD = 2(x + 2)$ cm, and $BC = x - 1$ cm. Find AD.
15. $AD = 2x + 3$ cm, $CD = 2x - 8$ cm, and $AB = x + 20$ cm. Find BC.

Rectangles

A *rectangle* is a parallelogram that contains a right angle. Since opposite angles are congruent in any parallelogram, the angle opposite the right angle is also a right angle. Since consecutive angles are supplementary in any parallelogram, the consecutive angles in the rectangle are also right angles, and so the rectangle has four right angles. Every rectangle is a parallelogram and has all the properties of a parallelogram. Its opposite sides are parallel and congruent, its opposite angles are congruent, its consecutive angles are supplementary, and its diagonals bisect each other. In addition, the diagonals of a rectangle are congruent.

Rhombuses

A *rhombus* is a parallelogram with four congruent sides. A rhombus has all the properties of a parallelogram. In addition, the diagonals of a rhombus are perpendicular, and the diagonals bisect the angles at the vertices they connect.

Squares

A *square* is a quadrilateral that is both a rectangle and a rhombus. Squares are parallelograms that have four congruent sides and four right angles. They have all the properties of parallelograms, all the properties of rectangles, and all the properties of rhombuses.

EXERCISE

6·6

Classify each quadrilateral as a parallelogram, rectangle, rhombus, or square.

1. 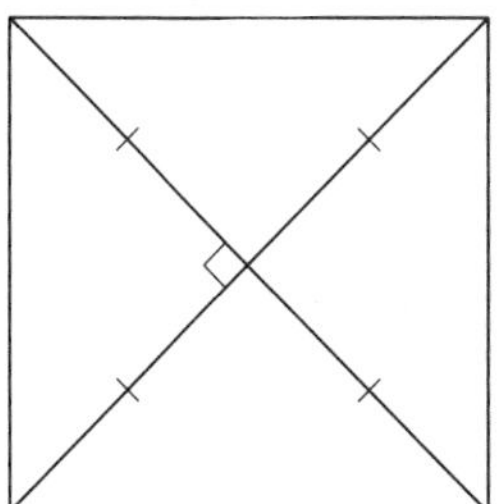

2. 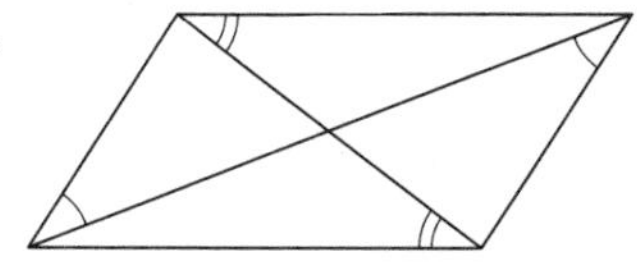

3. 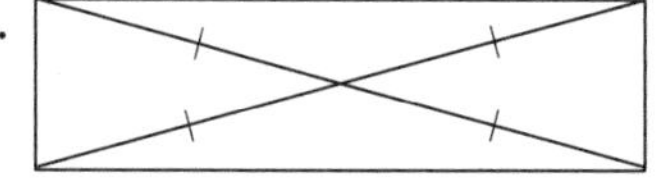

4. 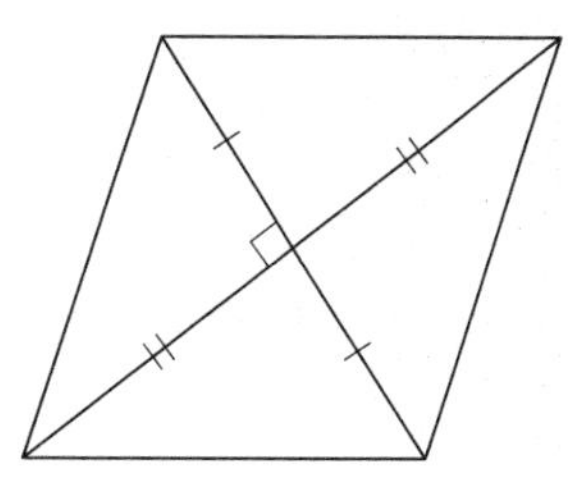

5. 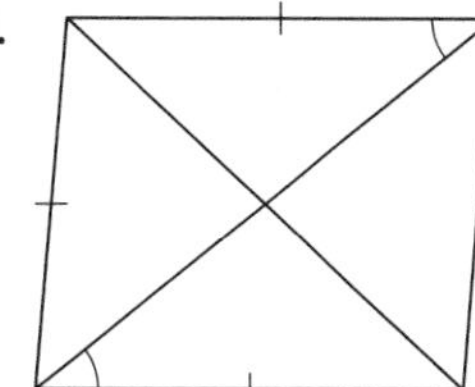

Find the length of the indicated segment.

6. $ABCD$ is a square with $AB = 12$ cm. Find the length of diagonal AC.
7. $PQRS$ is a rhombus with diagonals PR and QS intersecting at T. $QR = 13$ in and $QT = 5$ in. Find RT.
8. $WXYZ$ is a rectangle. $WX = 4$ m and $XY = 3$ m. Find XZ.
9. $JKLM$ is a rhombus with diagonals JL and KM intersecting at N. $JL = 18$ cm and $KL = 15$ cm. Find MK.
10. $ABCD$ is a rectangle with diagonal $AC = 26$ m. $AB = x$ and $BC = 2(x + 2)$. Find CD.

Find the measurement of the indicated angle.

11. $PQRS$ is a rectangle with diagonal PR. $m\angle PRS = 28°$. Find $m\angle SPR$.
12. $ABCD$ is a rhombus with diagonals AC and DB intersecting at E. Find $m\angle AED$.
13. $JKLM$ is a square with diagonals JL and KM intersecting at N. Find $m\angle NJK$.
14. $WXYZ$ is a rhombus with diagonal XZ. $m\angle XZY = 53°$. Find $m\angle ZWX$.
15. $JKLM$ is a rhombus with diagonals JL and MK intersecting at N. $m\angle JKN = 9x - 7°$ and $m\angle MJN = 10x + 2°$. Find $m\angle MLK$.

Proofs about quadrilaterals

Some problems will ask you to prove that a figure is a certain type of quadrilateral, and others will tell you that the figure is a certain type of quadrilateral and ask you to use the properties of that quadrilateral to prove something else. Whichever way the question is posed, it's important that you have a clearly organized understanding of all the different quadrilaterals, their definitions, and their properties.

Don't forget that any ideas you've learned before can be used in new proofs. Congruent triangles are always a powerful tool, and you can expect them to pop up again and again.

1. Given $\angle SPQ \cong \angle SRQ$ and $\overline{PR}$ bisects $\angle SPQ$ and $\angle SRQ$. Prove *PQRS* is a rhombus.

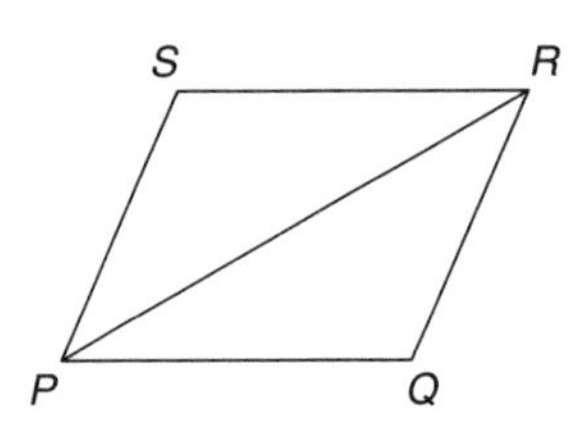

2. Given *ACEF* is a parallelogram and *ABDF* is a rectangle. Prove $\overline{BC} \cong \overline{DE}$.

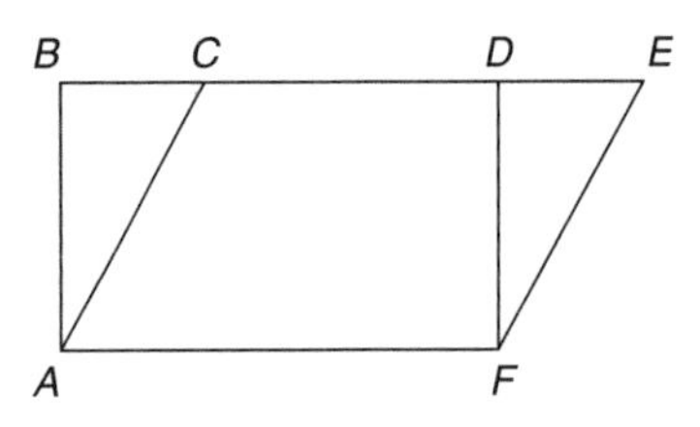

3. Given $\overline{VX} \cong \overline{VY}$ and $\overline{WV} \cong \overline{VZ}$. Prove *WXYZ* is an isosceles trapezoid.

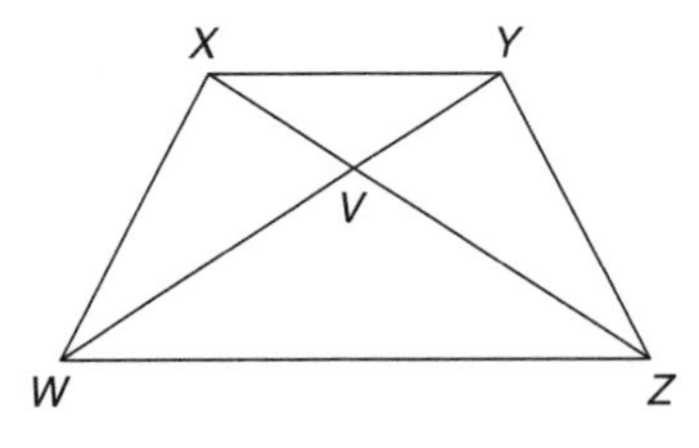

4. Given *PQRS* is a square, $\overline{RW} \perp \overline{TS}$, $\overline{PV} \perp \overline{QU}$, and $\overline{RW} \cong \overline{PV}$. Prove *QUST* is a parallelogram.

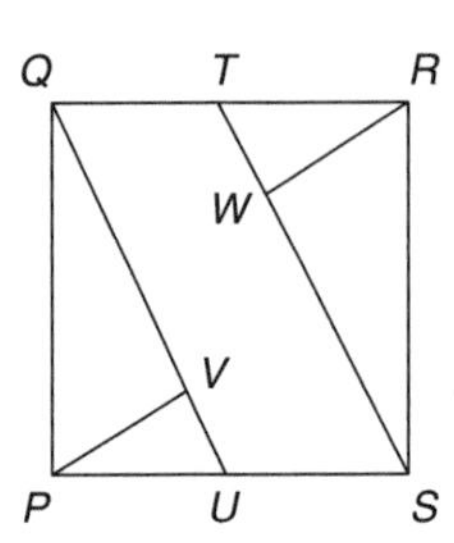

5. Given *RTVW* is a trapezoid, $\overline{RT} \cong \overline{TU}$, and $\overline{ST} \cong \overline{TV}$. Prove *RSVW* is a parallelogram.

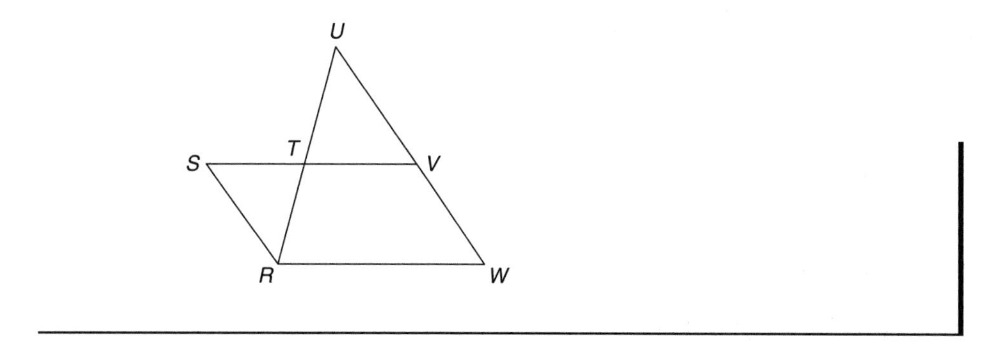

Similarity

Congruent polygons are the same size and shape, but similar polygons are the same shape but not necessarily the same size. If you enlarge a picture, you want to do it in a way that doesn't distort the image. That requires that the dimensions be increased proportionally, which means the enlarged picture is similar to the original.

Ratio and proportion

A *ratio* is a comparison of two numbers by division. The relationship between 10 and 5 or between 26 and 13, for example, can be expressed as a ratio: 10:5 or $\frac{26}{13}$. Both of these ratios are equal to 2:1.

Two equal ratios form a *proportion*. In a proportion such as 10:5 = 2:1, the numbers on the ends, 10 and 1, are called the *extremes*, and the numbers in the middle, 5 and 2, are called the *means*. When the ratios are written as fractions, the proportion is

$$\frac{\text{extreme}}{\text{mean}} = \frac{\text{mean}}{\text{extreme}}$$

If the same number shows up in both of the mean positions, we say it is the geometric mean between the two extremes, or the mean proportional of the two extremes. If $\frac{4}{8} = \frac{8}{16}$, then 8 is the geometric mean between 4 and 16, or the mean proportional of 4 and 16.

In any proportion, the product of the means is equal to the product of the extremes. If $\frac{4}{8} = \frac{8}{16}$, then $8 \cdot 8 = 4 \cdot 16$. This means it is possible to use cross-multiplication to solve for the missing term of a proportion. If $\frac{3}{5} = \frac{x}{20}$, the product of the means, $5x$, is equal to the product of the extremes, 60. You can solve $5x = 60$ to find out that x should be 12.

EXERCISE

7·1

Solve each proportion to find the value of the variable.

1. $\frac{5}{3} = \frac{7}{x}$
2. $\frac{w}{5} = \frac{6}{2}$
3. $\frac{9}{15} = \frac{15}{x}$
4. $\frac{2.5}{18} = \frac{30}{x}$

5. $\frac{4}{x+5}=\frac{3}{x}$

6. $\frac{x-7}{2}=\frac{2x-3}{5}$

7. $\frac{x}{4}=\frac{16}{x}$

8. $\frac{3}{x-4}=\frac{x+4}{3}$

9. $\frac{5}{x-12}=\frac{x+12}{5}$

10. $\frac{7}{x-3}=\frac{x}{4}$

Similar polygons

Two polygons are *similar* if their vertices can be matched in such a way that corresponding angles are congruent and corresponding sides are in proportion. The similarity statement tells which vertices correspond and therefore which angles are congruent and which sides are proportional. If quadrilateral $ABCD \sim PMNO$, $\angle A \cong \angle P$, $\angle B \cong \angle M$, $\angle C \cong \angle N$, and $\angle D \cong \angle O$. Sides are also proportional, and the relationship is

$$\frac{AB}{PM}=\frac{BC}{MN}=\frac{CD}{NO}=\frac{AD}{PO}$$

The ratio of the sides is called the *scale factor.*

Write the proportion that relates the corresponding sides of the similar polygons.

1. Pentagon *ABCDE* ~ pentagon *VWXYZ*
2. Quadrilateral *STEP* ~ quadrilateral *MAIL*
3. $\Delta CAT \sim \Delta DOG$
4. Rectangle *HAND* ~ rectangle *FORK*
5. Hexagon *BRANCH* ~ hexagon *MISTED*

Proving triangles similar

The primary method of proving triangles similar is to prove that two angles of one triangle are congruent to the corresponding angles of the other triangle. This is abbreviated AA. It isn't necessary to prove all three pairs of corresponding angles congruent, because if two angles of one triangle are congruent to two angles of another triangle, the third angles are congruent as well.

It is possible to prove two triangles similar by showing that all three pairs of corresponding sides are in proportion. The abbreviation for this is SSS, but you must remember that it is not the same SSS theorem you learned in congruent triangles. This one says the sides are in proportion, not that they are congruent.

There is also an SAS postulate for similar triangles. It says that the two triangles will be similar if two pairs of corresponding sides are in proportion and the angles included between those pairs of sides are congruent.

EXERCISE

7·3

If there is enough information to conclude that the triangles are similar, give the postulate or theorem that supports the conclusion. If not, write "Cannot be determined."

1. 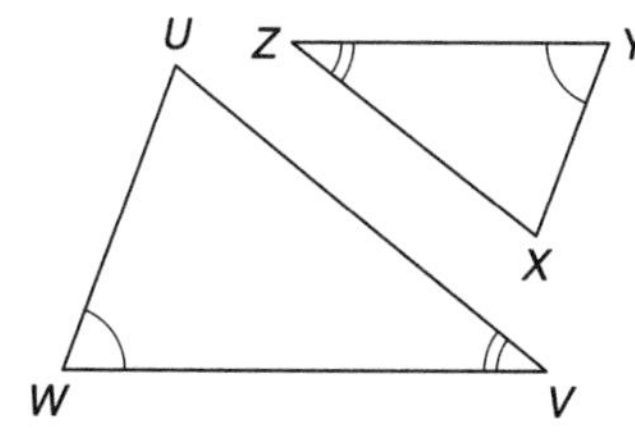

2. 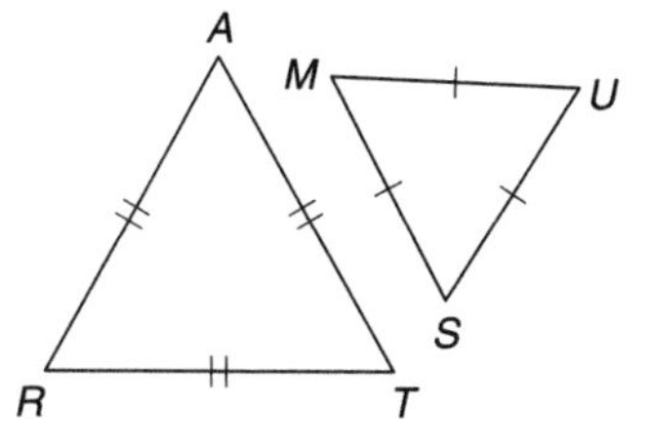

3. 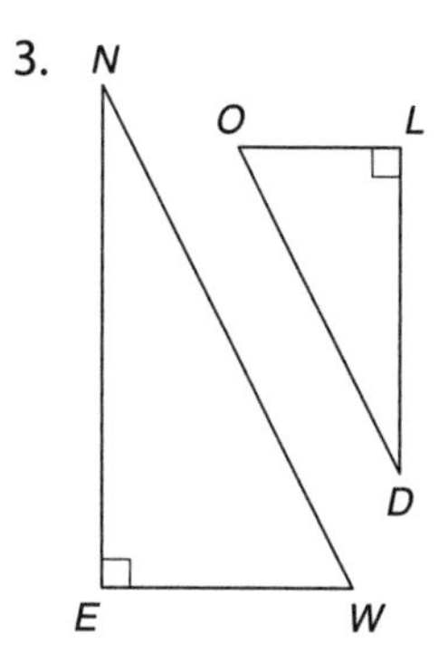

4. 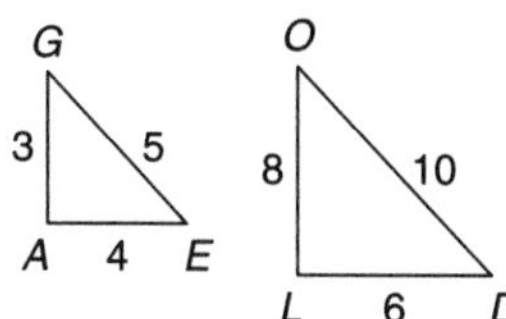

5. 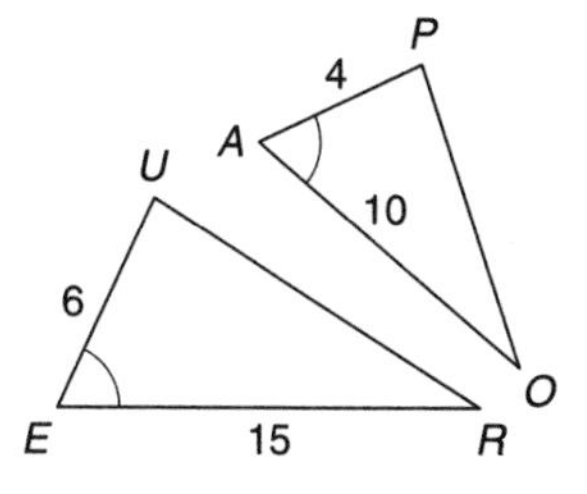

6. 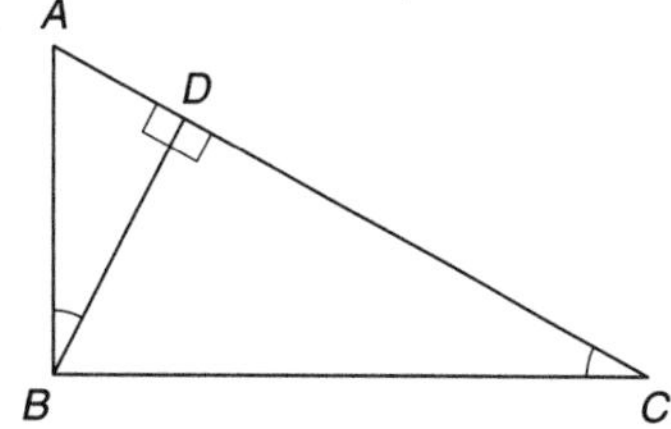

7. 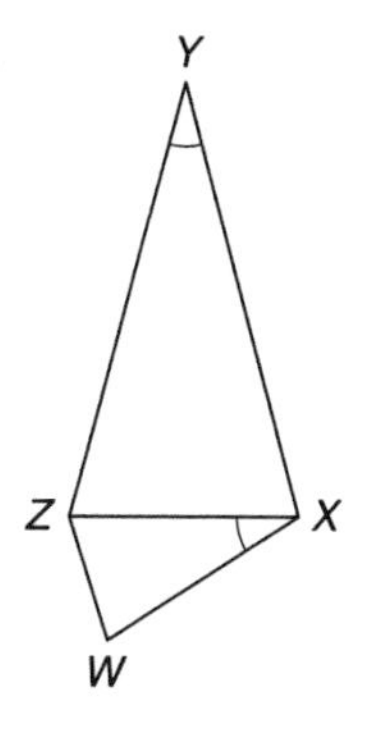

8. 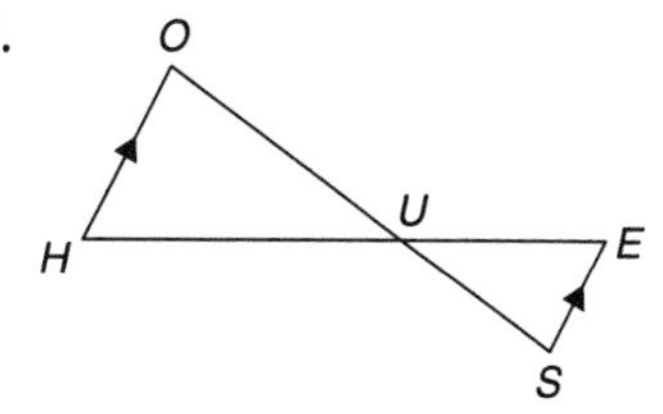

9. 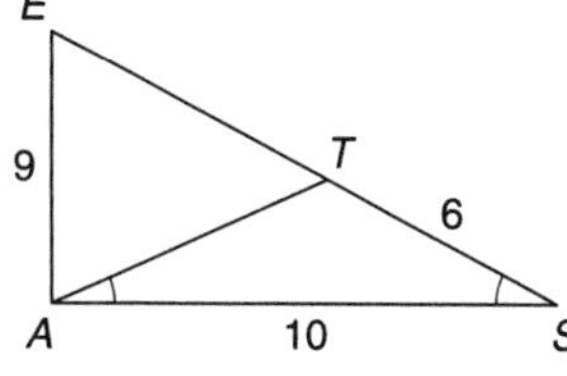

10. 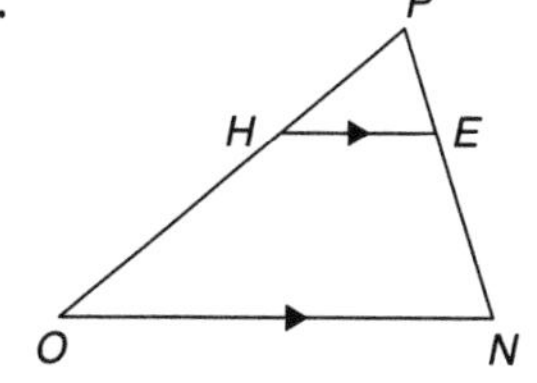

Proving segments proportional

While some problems will ask you to prove a pair of triangles similar, others will ask you to prove a proportion. The key lies in recognizing which pair of triangles, if shown to be similar, will yield the correct proportion. Given a proportion such as

$$\frac{XZ}{BC} = \frac{XY}{AB}$$

Identify the two triangles first. In this case, you want to work with ΔXYZ and ΔABC, but before writing a similarity statement, you must determine the correspondence. Look for the vertex named twice. Now X appears in both XZ and XY, and B appears in both BC and AB, so X corresponds to B. And XZ corresponds to BC, which means Z corresponds to C, and so Y corresponds to A. The similarity statement therefore is $\Delta XYZ \sim \Delta BAC$.

EXERCISE 7·4

Write the proportion that follows from each similarity statement.

1. $\Delta ABC \sim \Delta XYZ$
2. $\Delta RST \sim \Delta FED$
3. $\Delta PQR \sim \Delta VXW$
4. $\Delta MLN \sim \Delta LJK$
5. $\Delta ZXY \sim \Delta BCA$

Write a similarity statement that tells which pair of triangles should be proved similar in order to prove the specified proportion.

6. $\frac{AB}{XY} = \frac{BC}{YZ}$
7. $\frac{RS}{MO} = \frac{RT}{NO}$
8. $\frac{PQ}{DE} = \frac{PR}{EF}$
9. $\frac{BC}{TS} = \frac{AB}{RT}$
10. $\frac{XZ}{PQ} = \frac{YZ}{QR}$

Find the length of the specified segment.

11. $\Delta ABC \sim \Delta XYZ$, $AB = 21$ cm, $BC = 54$ cm, $XY = 7$ cm. Find YZ.
12. $\Delta DEF \sim \Delta CAT$, $DE = 65$ in, $EF = 45$ in, $CA = 13$ in. Find AT.
13. $\Delta GHI \sim \Delta ARM$, $GH = 9$ ft, $GI = 8$ ft, $AR = 12$ ft. Find AM.
14. $\Delta JKL \sim \Delta DOG$, $JK = 17$ m, $JL = 25$ m, $DG = 30$ m. Find DO.
15. $\Delta MNO \sim \Delta LEG$, $MN = x-3$, $NO = 3$, $EG = 21$, $LE = 2x+4$. Find LE.
16. $\Delta PQR \sim \Delta TOE$, $PQ = 3x+1$, $PR = 5$, $TE = 30$, $TO = 21x$. Find TO.
17. $\Delta STU \sim \Delta EAR$, $ST = 69$, $TU = 27$, $AR = x-2$, $EA = 2x+1$. Find EA.
18. $\Delta VWX \sim \Delta LIP$, $VW = 2x+6$, $VX = 6$, $LP = 7$, $LI = 3x-5$. Find LI.
19. $\Delta RST \sim \Delta CAP$, $RS = x+5$, $RT = 5x-1$, $CA = 1$, $CP = x-5$. Find CP.
20. $\Delta LMN \sim \Delta RAT$, $LM = 2x+5$, $MN = x+4$, $RA = x$, $AT = x-2$. Find AT.

Side-splitter theorem

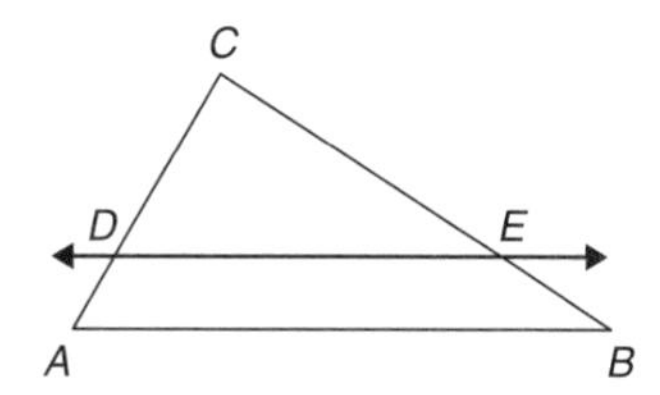

Figure 7.1 Line *DE* splits sides *CA* and *CB*.

A line parallel to one side of a triangle that intersects the other two sides divides the sides proportionally (see Figure 7.1). The triangle cut off by this line is similar to the original triangle. This can be proved quickly by proving $\angle CDE \cong \angle A$ and $\angle CED \cong \angle B$. Because the triangles are similar,

$$\frac{CD}{CA} = \frac{CE}{CB} = \frac{DE}{AB}$$

With a little bit of algebra it is possible to show that the line divides the sides proportionally. Start with

$$\frac{CD}{CA} = \frac{CE}{CB}$$

and cross-multiply to get $CD \cdot CB = CA \cdot CE$. Subtract $CD \cdot CE$ from both sides, and factor to get $CD(CB - CE) = CE(CA - CD)$. But $CB - CE$ is EB and $CA - CD$ is DA, so the equation becomes $CD(EB) = CE(DA)$. Divide both sides by $EB \cdot DA$ and you have

$$\frac{CD}{DA} = \frac{CE}{EB}$$

This tells you that the line parallel to one side of the triangle divides the other two sides proportionally.

Be careful not to mix up pieces of one relationship with pieces of the other. Remember that DE/AB is equal to CD/CA and CE/CB, all sides of the triangles, but not to CD/DA or CE/EB. Those last two ratios refer to the parts of the broken sides. If the problem involves the parallel segments, use the ratio of the two triangles, not the ratio of the pieces of the sides.

In $\triangle XYZ$, $\overline{VW} \parallel \overline{XZ}$. Complete each proportion.

1. $\frac{YV}{VX} = \frac{YW}{?}$
2. $\frac{YV}{YX} = \frac{YW}{?}$
3. $\frac{YV}{?} = \frac{VW}{XZ}$
4. $\frac{?}{WZ} = \frac{YV}{VX}$
5. $\frac{VX}{?} = \frac{YV}{YW}$

Find the length of the indicated segment.

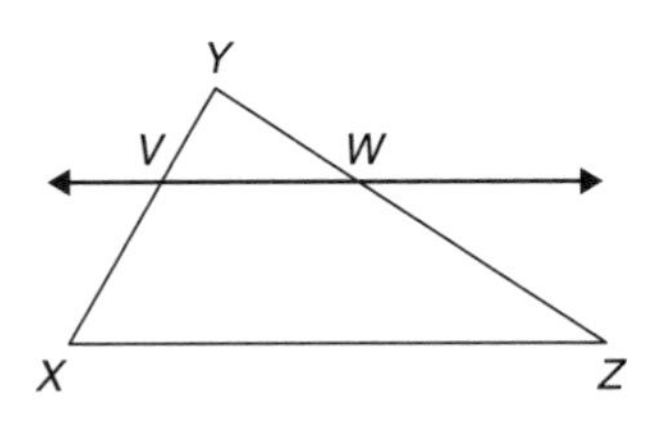

6. $VY = 5$, $WY = 15$, $XV = 8$. Find WZ.
7. $XV = 7$, $WZ = 28$, $WY = 12$. Find YV.
8. $YX = 17$, $YV = 8$, $YW = WZ - 2$. Find WZ.
9. $VY = x - 7$, $VW = x + 7$, $XV = x + 6$, $XZ = 4x + 1$. Find XZ.
10. $VY = x + 12$, $YW = x - 3$, $YZ = 4x - 3$, $XY = 228 - 5x$. Find WZ.

Dividing transversals proportionally

When three or more parallel lines are intersected by two transversals, the parallel lines divide the transversals proportionally. In Figure 7.2, $a/b = c/b$. The proportional relationship is an extension of the side-splitter theorem.

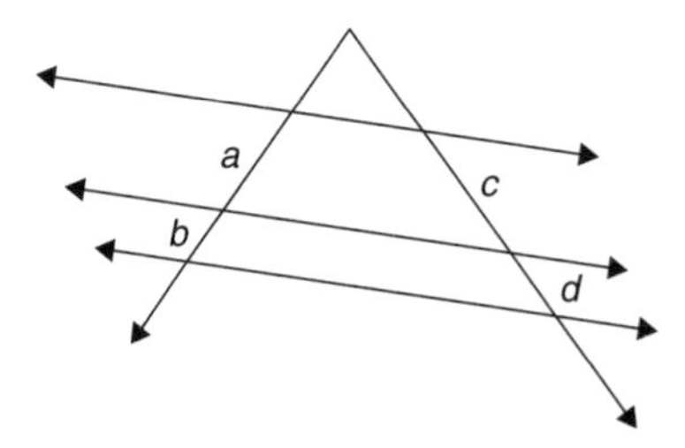

Figure 7.2 Three parallel lines with two transversals.

$\overline{AD} \parallel \overline{BE} \parallel \overline{CF}$. *Find the length of the indicated segment.*

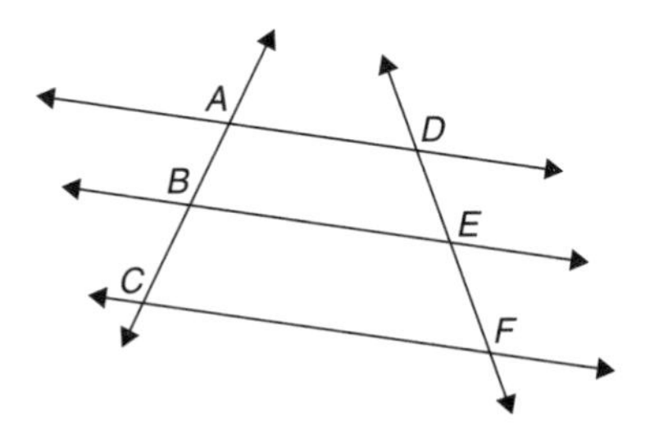

1. $AB = 8$, $BC = 4$, $DE = 12$. Find EF.
2. $AB = 3.5$, $BC = 8.5$, $DE = 10.5$. Find EF.
3. $AB = 51$, $DE = 17$, $EF = 33$. Find BC.
4. $BC = 19$, $AC = 34$, $ED = 45$. Find EF.
5. $AB = 18.3$, $AC = 40$, $EF = 86.8$. Find ED.
6. $DE = 5$, $DF = AC + 9$, $BC = 4$. Find AC.
7. $EF = 14x - 1$, $BC = 3x - 2$, $AB = 2x + 1$, $DE = 10x + 5$. Find DE.
8. $AC = 20$, $DF = 9x + 3$, $AB = x$, $DE = 12$. Find DF.
9. $AC = 4x - 5$, $DF = 18x + 9$, $EF = 12x - 9$, $BC = 2x + 5$. Find DE.
10. $AB \cdot EF = 144$, $DE = BC$. Find BC.

Similar-triangle proofs

Similar-triangle proofs are usually straightforward problems, but it's always a good idea *to plan* before you begin. Often they won't say, "prove these triangles are similar," but rather will ask you to prove that lengths of segments are in proportion, or that the product of the lengths of two segments equals the product of the lengths of two other segments. You need to take the time to work backward from what you're asked to prove and to determine which pair of triangles will give you the proportion you need.

Make a plan for each proof.

1. 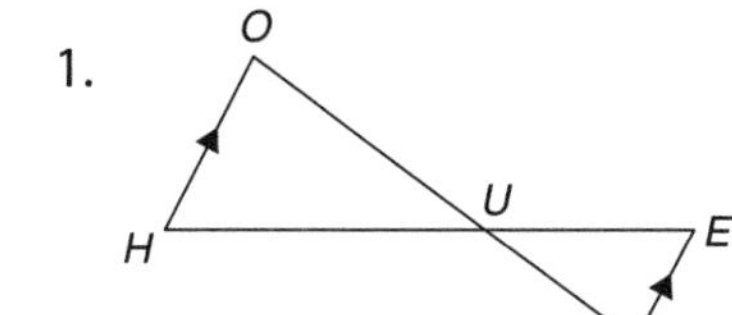

Given $\overline{HO} \parallel \overline{SE}$. Prove $\frac{OU}{US} = \frac{HU}{UE}$.

2. Given $\angle XWY \cong \angle XZV$. Prove $\frac{VX}{YX} = \frac{XZ}{YW}$.

3. 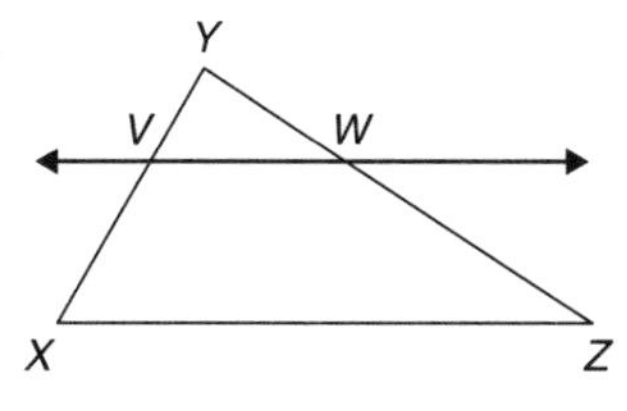

Given $\overleftrightarrow{VW} \parallel \overline{XZ}$. Prove $VY \cdot WZ = YW \cdot VX$.

4. 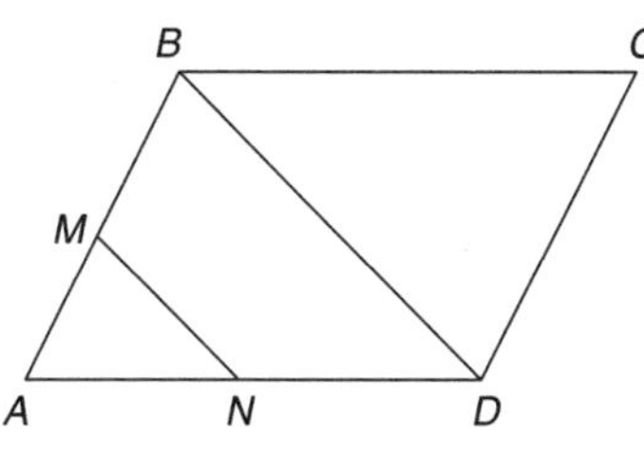

Given $ABCD$ is a parallelogram with diagonal $\overline{BD}$. M is the midpoint of $\overline{AB}$ and N is the midpoint of $\overline{AD}$. Prove $MN = \frac{1}{2}BD$.

5. 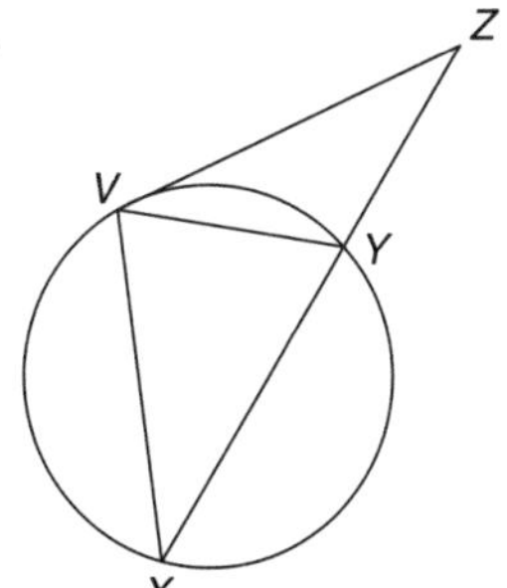

Given $\angle X \cong \angle ZVY$. Prove $YZ \cdot XZ = VZ^2$.

Right triangles

·8·

A *right triangle* is a triangle that contains one right angle. The other two angles are acute and complementary. The two sides that form the right angle are called *legs*, and the side opposite the right angle, the longest side, is called the *hypotenuse*.

The Pythagorean theorem

The *Pythagorean theorem* relates the lengths of the sides of a right triangle. In any right triangle, the square of the hypotenuse is equal to the sum of the squares of the other two sides. If a and b are the lengths of the legs of a right triangle and c is the length of the hypotenuse, then $a^2+b^2=c^2$. If the lengths of two sides are known, the Pythagorean theorem can be used to find the length of the third side of the right triangle.

Because square roots are involved in finding the lengths of the sides—$c=\sqrt{a^2+b^2}$ or $a=\sqrt{c^2-b^2}$—the lengths often work out to irrational numbers. You can leave these in simplest radical form or use a calculator to get a decimal approximation. A set of three integers that can be the sides of a right triangle and that fit the Pythagorean theorem are called a *Pythagorean triple*. Common Pythagorean triples are 3, 4, 5 and 5, 12, 13, as well as their multiples.

If a *and* b *are the lengths of the legs of a right triangle and* c *is the length of the hypotenuse, find the missing length.*

1. $a=3, b=4, c=?$
2. $a=5, b=?, c=13$
3. $a=?, b=32, c=40$
4. $a=7, b=9, c=?$
5. $a=6, b=?, c=\sqrt{41}$
6. $a=\sqrt{2}, b=\sqrt{3}, c=?$
7. $a=7, b=7, c=?$
8. $a=11, b=?, c=22$
9. $a=30, b=?, c=78$
10. $a=19, b=21, c=?$

Converse of the Pythagorean theorem

The $a^2+b^2=c^2$ relationship among the sides of a right triangle applies only to right triangles, so you can use it to determine whether a triangle is a right triangle. Call the longest side of the triangle c and the other two sides a and b, and then test to see if they fit the $a^2+b^2=c^2$ relationship. If they do, the triangle is a right triangle.

If the sides don't fit the Pythagorean relationship, there's still more information you can gather. If $a^2+b^2<c^2$, the triangle is obtuse; but if $a^2+b^2>c^2$, the triangle is acute. If $c^2=a^2+b^2$, the hypotenuse c fits in to the L shape made when the two legs are perpendicular. If $c^2>a^2+b^2$, the c side is too big for that L shape, and the legs have to bend out into an obtuse angle to make room. If $c^2<a^2+b^2$, the c side is too small, and the other two sides close into an acute angle.

EXERCISE
8·2

If the numbers given can be the lengths of the sides of a triangle, determine whether the triangle is right, acute, or obtuse.

1. $a=3, b=5, c=6$
2. $a=4, b=7, c=8$
3. $a=5, b=8, c=9$
4. $a=11, b=13, c=18$
5. $a=12, b=15, c=19$
6. $a=12, b=16, c=20$
7. $a=13, b=20, c=24$
8. $a=15, b=25, c=5\sqrt{34}$
9. $a=18, b=25, c=15\sqrt{5}$
10. $a=20,\ b=30, c=10\sqrt{13}$

Special right triangles

You'll find that it's helpful to be well acquainted with two families of special right triangles. In any isosceles right triangle, the two acute angles both measure 45° and the legs are the same length. So in the $a^2+b^2=c^2$ relationship, you can replace b with a, and you find that $a^2+a^2=2a^2=c^2$ and $c=a\sqrt{2}$. This tells you that in an isosceles right triangle, also known as a 45°-45°-90° triangle, the hypotenuse is equal to the leg times the square root of 2. Think "side-side-side radical 2." You can turn that around and say the leg is the hypotenuse divided by the square root of 2, which simplifies to $a\sqrt{2}/2$.

The other family of special right triangles actually starts with an equilateral triangle, which has three congruent sides and three 60° angles. When you draw an altitude, you divide the equilateral triangle into two congruent right triangles, each with angles measuring 30°, 60°, and 90°. In these 30°-60°-90° right triangles, the hypotenuse is a side of the original equilateral triangle, and the side opposite the 30° angle is one-half of that. You can use the Pythagorean theorem to show that the remaining side, opposite the 60° angle, is one-half the hypotenuse times the square root of 3. Pull all that information together, and you know that in a 30°-60°-90° triangle, the side opposite the 30° angle is one-half the hypotenuse and the side opposite the 60° angle is one-half the hypotenuse times radical 3. Think "hypotenuse, one-half the hypotenuse, one-half the hypotenuse radical 3."

ΔABC is a right triangle with $\overline{AB}\perp\overline{BC}$. Given the measure of one of the acute angles of the right triangle and the length of one side, find the lengths of the other two sides.

1. $\angle A=30°, AB=5\sqrt{3}$
2. $\angle A=60°, BC=7\sqrt{3}$
3. $\angle A=30°, AC=18$
4. $\angle A=60°, AB=17$

5. $\angle A = 30°$, $BC = 23$
6. $\angle A = 45°$, $AC = 8\sqrt{2}$
7. $\angle A = 45°$, $AB = 5$
8. $\angle A = 45°$, $BC = 3$
9. $\angle A = 45°$, $AC = 12$
10. $\angle A = 45°$, $AB = 4\sqrt{2}$
11. $\angle A = 30°$, $BC = 5\sqrt{3}$
12. $\angle A = 45°$, $AC = 9$
13. $\angle A = 60°$, $AB = 2\sqrt{5}$
14. $\angle A = 45°$, $BC = 2\sqrt{3}$
15. $\angle A = 30°$, $AC = 25$
16. $\angle A = 45°$, $AB = 11.2$
17. $\angle A = 60°$, $BC = 12$
18. $\angle A = 45°$, $AC = 8\sqrt{6}$
19. $\angle A = 30°$, $AB = 19$
20. $\angle A = 60°$, $AC = 11\sqrt{5}$

Altitude to the hypotenuse

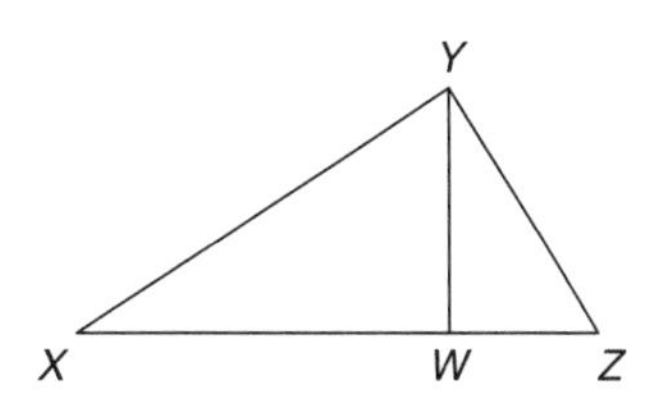

Figure 8.1

In a right triangle, you can look at the legs, the two perpendicular sides, as being a base and an altitude (see Figure 8.1). When an altitude is drawn from the right-angle vertex to the hypotenuse of a right triangle, however, the altitude divides the right triangle into two smaller right triangles, each of which is similar to the original. As a result, the two small triangles are similar to each other. $\Delta XYZ \sim \Delta XWY \sim \Delta YWZ$. It's important to pay close attention to the correspondence in these similarity statements. It's not the correspondence you might expect from looking at the diagram.

The similarity statement $\Delta XYZ \sim \Delta XWY$ leads to the proportion $\frac{XY}{XW} = \frac{YZ}{WY} = \frac{XZ}{XY}$.

The similarity statement $\Delta XYZ \sim \Delta YWZ$ leads to the proportion $\frac{XY}{YW} = \frac{YZ}{WZ} = \frac{XZ}{YZ}$.

The similarity statement $\Delta XWY \sim \Delta YWZ$ leads to the proportion $\frac{XW}{YW} = \frac{WY}{WZ} = \frac{XY}{YZ}$.

Altitude as geometric mean

Because the two small triangles are similar, $\Delta XWY \sim \Delta YWZ$, their sides are in proportion:

$$\frac{XW}{YW} = \frac{WY}{WZ} = \frac{XY}{YZ}$$

The first part of this proportion, $XW/YW = WY/WZ$, shows that when the altitude is drawn to the hypotenuse of a right triangle, the length of the altitude is the geometric mean between the two parts of the hypotenuse.

If the altitude to the hypotenuse of a right triangle divides the hypotenuse into two segments that are 4 in and 9 in long, then the length of the altitude can be found by solving $4/x = x/9$ for x. If you cross-multiply, you find that $x^2 = 36$, so $x = 6$.

Altitude $\overline{YW}$ is drawn to hypotenuse $\overline{XZ}$ of right triangle XYZ. Find the length of the indicated segment.

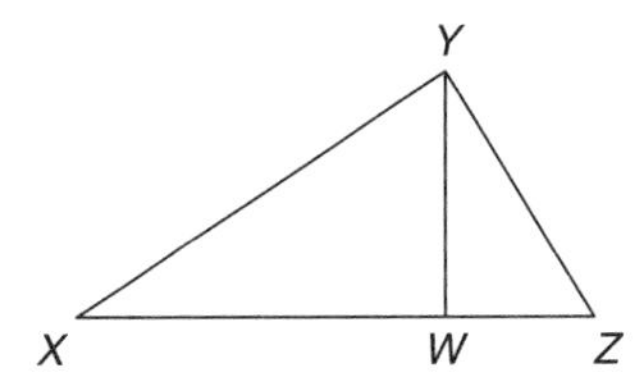

1. $XW = 4$, $WZ = 2.25$. Find YW.
2. $XW = 4$, $WZ = 12$. Find YW.
3. $YW = 6$, $WZ = 4$. Find XW.
4. $XW = 4$, $WZ = 16$. Find YW.
5. $XW = 5$, $XZ = 15$. Find YW.
6. $XZ = 12$, $XW = YW$. Find YW.
7. $XW = YW - 5$, $WZ = 2YW$. Find YW.
8. $XW = x + 1$, $WZ = x - 1$, $YW = x - 3$. Find XZ.
9. $XW = 2x - 1$, $WZ = x + 3$, $YW = x + 7$. Find XW.
10. $XW = 3x - 1$, $XZ = 4x + 2$, $YW = 2x + 1$. Find YW.

Leg as a geometric mean

Looking at the similarity relationship between a small triangle and the larger, original right triangle, you can see that each leg of the original right triangle is a geometric mean between the hypotenuse and the portion of the hypotenuse nearest that leg. The proportion

$$\frac{XY}{XW} = \frac{YZ}{WY} = \frac{XZ}{XY}$$

can be rewritten (and the middle ratio ignored) to give $XW/XY = XY/XZ$. The length of leg $\overline{XY}$ is the geometric mean between hypotenuse $\overline{XZ}$ and $\overline{XW}$, the part of the hypotenuse near $\overline{XY}$. In the same way,

$$\frac{XY}{YW} = \frac{YZ}{WZ} = \frac{XZ}{YZ}$$

can be inverted and the first ratio ignored to produce $WZ/YZ = YZ/XZ$. The length of side $\overline{YZ}$ is the geometric mean between hypotenuse $\overline{XZ}$ and $\overline{WZ}$, the part of the hypotenuse near $\overline{YZ}$.

If the altitude is drawn to the hypotenuse of a 3 cm–4 cm–5 cm right triangle, it divides the hypotenuse into two segments with lengths that can be represented as x and $5 - x$. Let's say the segment with length x is near the 3-cm side and the segment with length $5 - x$ is near the 4-cm side. You can write two proportions:

$$\frac{x}{3} = \frac{3}{5} \quad \text{and} \quad \frac{5-x}{4} = \frac{4}{5}$$

The first one says the 3-cm leg is the geometric mean between the part of the hypotenuse with length x and the whole hypotenuse with length 5 cm. The second proportion makes a similar statement about the 4-cm leg, the segment of the hypotenuse of length $5 - x$ and the whole hypotenuse. Solving

$$\frac{x}{3} = \frac{3}{5}$$

gives you $5x = 9$ and $x = 1.8$ cm. Solving

$$\frac{5-x}{4} = \frac{4}{5}$$

gives you $25 - 5x = 16$, which also produces $x = 1.8$ cm. The altitude to the hypotenuse of a 3-4-5 right triangle divides the hypotenuse into two segments, one 1.8 cm long and the other 3.2 cm long.

Altitude $\overline{YW}$ is drawn to hypotenuse $\overline{XZ}$ of right triangle XYZ. Find the length of the indicated segment.

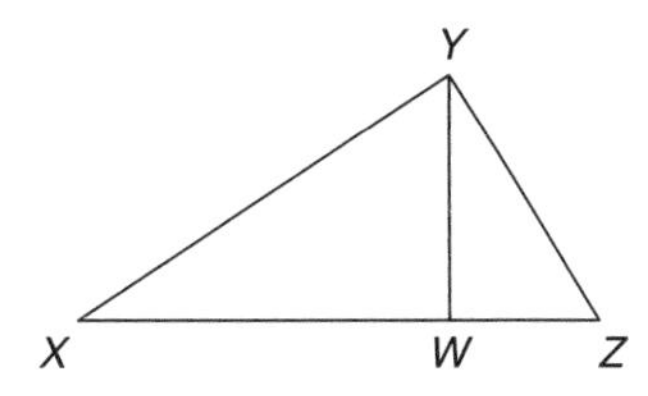

1. $XW = 4$, $XY = 6$. Find XZ.
2. $YZ = 10$, $XZ = 25$. Find WZ.
3. $XY = 12$, $XZ = 24$. Find XW.
4. $WZ = 4$, $YZ = 18$. Find XW.
5. $YZ = WZ + 1$, $XW = 3$. Find YZ.
6. $XY = 15$, $XW = 5x$, $WZ = 3x + 2$. Find XZ.
7. $XY = 8$, $WZ = 12$. Find XW.
8. $XY = 15$, $XW = 12$. Find WZ.
9. $WZ = x$, $XW = x + 5$, $YZ = 5$. Find XZ.
10. $YZ = 6$, $WZ = 3x + 1$, $XW = 2x + 3$. Find XZ.

Circles

·9·

A *circle* is the set of all points at a fixed distance from a given point. The given point is called the *center* of the circle, and the fixed distance is the *radius*. The word *radius* is also used to denote a line segment that connects the center to a point on the circle.

Lines and segments in circles

A *chord* is a line segment whose endpoints are points of the circle. A *diameter* is a chord that passes through the center of the circle, and it is the longest chord of the circle. A *secant* is a line that contains a chord and intersects the circle in two points. A *tangent* is a line that intersects the circle in exactly one point. The point at which the tangent touches the circle is called the *point of tangency*.

A tangent may touch two circles, each in a single point. In this case, it is called a *common tangent*. A common tangent is external if it does not intersect the line segment that connects the centers of the circles. A common internal tangent crosses the line segment containing the centers.

Two circles are tangent to each other if they share exactly one point. They may be internally or externally tangent. If the circles are internally tangent, one is inside the other. Concentric circles have the same center, but different radii. Concentric circles are not tangent because the center they have in common is not a point of the circle.

EXERCISE 9·1

Use the following diagram to fill each blank with one of the words below.

Center	Chord	Circle	Radius
Secant	Tangent	Diameter	

1. $\overleftrightarrow{AB}$ is a ______________ of circle C.
2. $\overline{JK}$ is a ______________ of circle O.
3. $\overline{FG}$ is a ______________ of circle O.
4. $\overline{CD}$ is a ______________ of circle C.
5. $\overleftrightarrow{LB}$ is a ______________ of circle O.

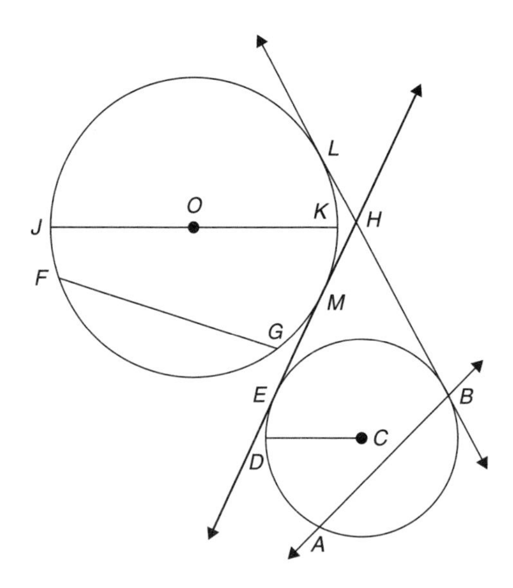

Identify each of the following in the diagram.

6. A diameter
7. A common internal tangent
8. A radius
9. A chord
10. A common external tangent

Angles in and around the circle

When radii, diameters, chords, secants, and tangents start to intersect one another, they form angles. The measure of an angle formed by such lines and segments intersecting depends on what segments intersect, where they intersect, and the size of the arcs, or portions of the circle, that they intercept.

When we write $\widehat{AB}$, we mean arc AB and when we write $m\widehat{AB}$, we mean the measure of arc AB, in degrees.

Vertex at the center: angle = arc

A *central angle* is an angle whose vertex is the center of the circle and whose sides are radii. The measure of a central angle is equal to the measure of its intercepted arc, and arcs, like angles, are measured in degrees. A full rotation is 360°, which is why we commonly say there is 360° in a circle. When we give the measurement of an arc in degrees, we're indicating what portion of a full rotation—what portion of a circle—it represents.

The length of the arc is a different matter. It depends upon the radius of the circle. A quarter rotation is 90° in any circle, but a very different length in a circle the size of a dime than in a circle the size of a ferris wheel.

Vertex on the circle: angle = $\frac{1}{2}$ arc

An *inscribed angle* is an angle whose vertex is a point on the circle and whose sides are chords. The measure of an inscribed angle is equal to one-half the measure of its intercepted arc.

An angle formed when a chord meets a tangent at the point of tangency is equal to one-half the measure of the intercepted arc. If the chord is a diameter, the angle is a right angle.

EXERCISE 9·2

In the circle with center O, $\overline{ON}$ and $\overline{OQ}$ are radii, and $\overline{PN}$ and $\overline{PQ}$ are chords. $\overleftrightarrow{RS}$ is tangent to the circle at P.

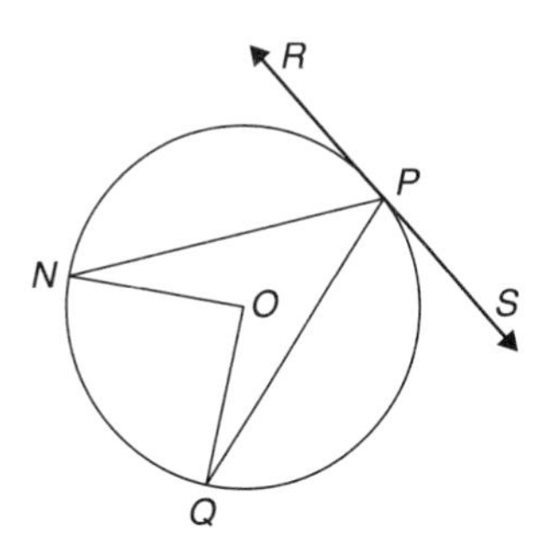

1. If $m\overset{\frown}{NQ} = 48°$, find $m\angle O$.
2. If $m\overset{\frown}{NQ} = 86°$, find $m\angle NPQ$.
3. If $m\overset{\frown}{NQ} = 40°$, find $m\angle O$.
4. If $m\overset{\frown}{NQ} = 66°$, find $m\angle NPQ$.
5. If $m\angle O = 45°$, find $m\overset{\frown}{NQ}$.
6. If $m\angle NPQ = 53°$, find $m\overset{\frown}{NQ}$.
7. If $m\angle O = 28°$, find $m\angle NPQ$.
8. If $m\overset{\frown}{PQN} = 230°$, find $m\angle SPN$.
9. If $m\overset{\frown}{PNQ} = 260°$, find $m\angle RPQ$.
10. If $m\overset{\frown}{PNQ} = 260°$, find $m\angle SPQ$.
11. If $m\overset{\frown}{NP} = 110°$, find $m\angle RPN$.
12. If $m\overset{\frown}{PQ} = 112°$, find $m\angle SPQ$.
13. If $m\angle NPQ = 82°$ and $\overset{\frown}{NP} \cong \overset{\frown}{PQ}$, find $m\angle SPQ$.
14. If $m\overset{\frown}{PNQ} = 242°$ and $m\angle NPQ = 47°$, find $m\angle RPN$.
15. If $m\overset{\frown}{NQ} = (2x+5)°$ and $m\angle O = (3x - 2)°$, find x.
16. If $m\overset{\frown}{NQ} = (160-6t)°$ and $m\angle NPQ = (3t + 8)°$, find t.
17. If $m\angle O = (6y-8)°$ and $m\angle NPQ = (4y - 23)°$, find y.
18. If $m\overset{\frown}{PQN} = (6p+14)°$ and $m\angle SPN = (193 - 3p)°$, find p.
19. If $m\overset{\frown}{NQ} = (3x+14)°$, $m\overset{\frown}{NP} = (4x-2)°$, and $m\overset{\frown}{PQ} = (5x)°$, find $m\angle NPQ$.
20. If $m\overset{\frown}{PNQ} = (x^2+3x+10)°$, $m\angle RPN = (5x + 13)°$, and $m\angle NPQ = (3x - 1)°$, find $m\angle RPN$.

Vertex inside the circle: angle = $\frac{1}{2}$ × sum of the arcs

When two chords intersect within the circle, vertical angles are formed. The vertical angles must be congruent, yet each angle in the pair of vertical angles may intercept a different-size arc. When two chords intersect in a circle, the measure of each of the angles in a pair of vertical angles is the average of the two intercepted arcs. In Figure 9.1,

$$m\angle 1 = m\angle 3 = \tfrac{1}{2}(m\overset{\frown}{AD} + m\overset{\frown}{BC}) \text{ and } m\angle 2 = m\angle 4 = \tfrac{1}{2}(m\overset{\frown}{AC} + m\overset{\frown}{BD}).$$

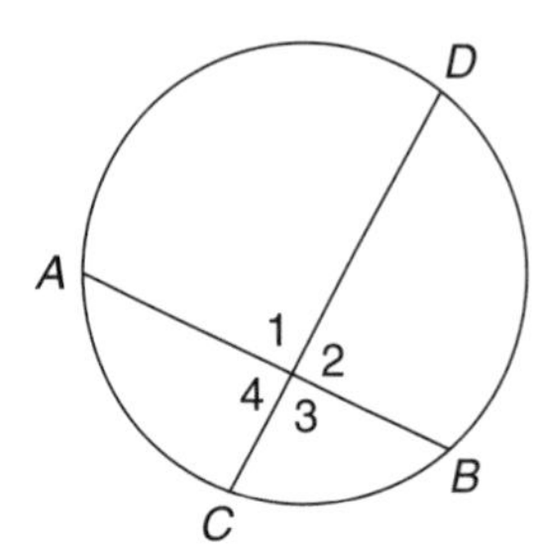

Figure 9.1

Use Figure 9.1 and the given information to find the specified measurement.

1. If $m\overset{\frown}{AD} = 96°$ and $m\overset{\frown}{BC} = 62°$, find $m\angle 1$.
2. If $m\overset{\frown}{AC} = 101°$ and $m\overset{\frown}{BD} = 93°$, find $m\angle 4$.
3. If $m\overset{\frown}{AD} = 118°$ and $m\overset{\frown}{BC} = 84°$, find $m\angle 3$.
4. If $m\overset{\frown}{AC} = 125°$ and $m\overset{\frown}{BD} = 63°$, find $m\angle 2$.
5. If $m\overset{\frown}{AC} = 211°$ and $m\angle 2 = 148°$, find $m\overset{\frown}{BD}$.
6. If $m\overset{\frown}{CB} = 113°$ and $m\angle 1 = 103°$, find $m\overset{\frown}{AD}$.
7. If $m\overset{\frown}{AC} = 87°$ and $m\overset{\frown}{BD} = 101°$, find $m\angle 1$.
8. If $m\overset{\frown}{AD} = (9x+9)°$, $m\overset{\frown}{BC} = (20x-9)°$, and $m\angle 1 = (15x - 6)°$, find $m\overset{\frown}{BC}$.
9. If $m\overset{\frown}{AC} = (13y+2)°$, $m\overset{\frown}{BD} = (25y)°$, and $m\angle 4 = (141 - y)°$, find $m\overset{\frown}{AC}$.
10. If $m\overset{\frown}{AD} = (26x-2)°$, $m\overset{\frown}{BC} = (x+6)°$, and $m\angle 1 = (14x - 2)°$, find $m\angle 4$.

Vertex outside the circle: angle = $\frac{1}{2}$ × difference of the arcs

When an angle is formed by two secants drawn from the same point outside the circle, the vertex of the angle is outside the circle and the sides of the angle cut across the circle. Every angle formed in this way intercepts two arcs, a small one when the sides first meet the circle and a larger one as the sides exit the circle. The measure of the angle is one-half the difference of the arcs.

Angles formed by a tangent and a secant, or by two tangents, from a single point outside the circle also have two intercepted arcs, but because the tangent just touches the circle at a single point, the intercepted arcs will meet each other. In the case of an angle formed by two tangents, the large and the small intercepted arcs together make up the whole circle.

Angles formed by two secants

When two secants are drawn to a circle from a single point, as in Figure 9.2, the angle formed is equal to one-half the difference of the far intercepted arc minus the near intercepted arc: $m\angle Q = \frac{1}{2}(m\overset{\frown}{PR} - m\overset{\frown}{ST})$.

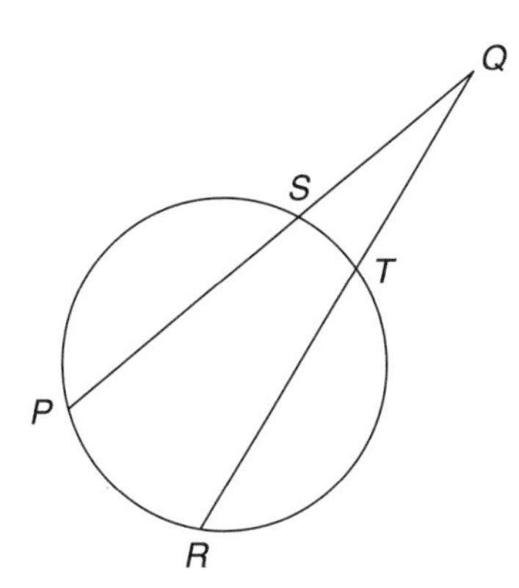

Figure 9.2 Formed by two secants, angle *Q* intercepts.

Angles formed by a tangent and a secant

An angle formed by a tangent and a secant from a single point is equal to one-half the difference of the two intercepted arcs. In Figure 9.3, notice that the two arcs meet at the point of tangency: $m\angle Z = \frac{1}{2}(m\overset{\frown}{VX} - m\overset{\frown}{VY})$.

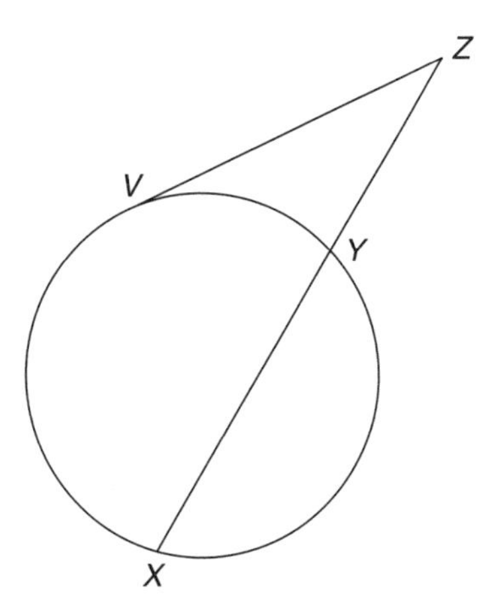

Figure 9.3 Angle *Z*, formed by a tangent and a secant.

Angles formed by two tangents

Figure 9.4 shows an angle formed by two tangents from a single point is equal to one-half the difference of the two intercepted arcs. The measures of the two intercepted arcs will add to 360°: $m\angle B = \frac{1}{2}(m\overset{\frown}{ADC} - m\overset{\frown}{AC})$.

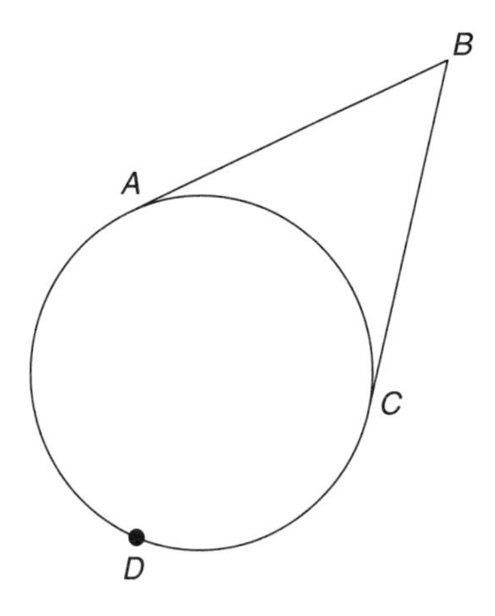

Figure 9.4 Angle B intercepts arc AC and arc ADC.

EXERCISE
9·4

Use the following diagram and the given information to find the specified measurements.

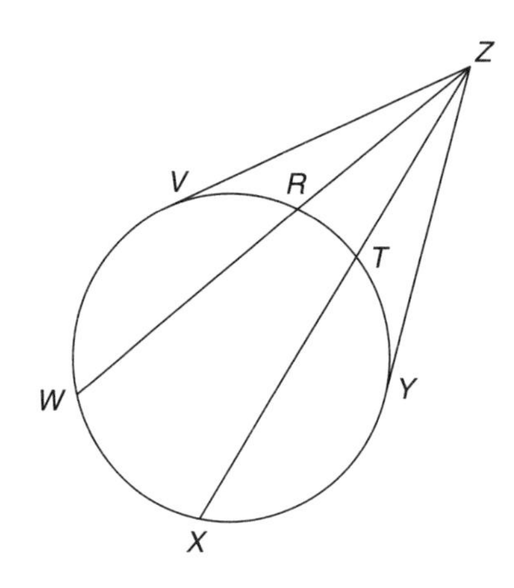

1. If $m\overset{\frown}{WX} = 85°$ and $m\overset{\frown}{RT} = 23°$, find $m\angle WZX$.
2. If $m\overset{\frown}{VW} = 93°$ and $m\overset{\frown}{VR} = 21°$, find $m\angle VZW$.
3. If $m\overset{\frown}{WXY} = 113°$ and $m\overset{\frown}{RY} = 39°$, find $m\angle WZY$.
4. If $m\overset{\frown}{VXY} = 285°$, find $m\angle VZY$.
5. If $m\overset{\frown}{VWX} = 157°$ and $m\angle VZX = 57°$, find $m\overset{\frown}{VRT}$.
6. If $m\overset{\frown}{WX} = 79°$ and $m\angle WZX = 33°$, find $m\overset{\frown}{RT}$.
7. If $m\overset{\frown}{WX} = (12x - 1)°$, $m\overset{\frown}{RT} = (2x + 5)°$, and $m\angle WZX = (81 - 7x)°$, find x.
8. If $m\overset{\frown}{VW} = (30 + 3x)°$, $m\overset{\frown}{VR} = (x - 6)°$, and $m\angle VZW = (2x - 1)°$, find x.
9. If $m\overset{\frown}{VXY} = (12x - 40)°$, $m\overset{\frown}{VTY} = (4x)°$, and $m\angle VZY = (3x + 5)°$, find $m\angle VZY$.
10. If $m\overset{\frown}{VWX} = (6x - 1)°$, $m\overset{\frown}{VRT} = (198 - 5x)°$, and $m\angle VZX = (x + 4)°$, find $m\overset{\frown}{VRT}$.

Line segments

In addition to forming angles, intersecting line segments in and around the circle divide one another in predictable ways. Some segments are congruent, while others are proportional, although the rules are usually given as products.

Two tangents

When two tangents are drawn to a circle from a single point, as in Figure 9.5, the tangent segments are congruent. A radius drawn to the point of tangency is perpendicular to the tangent.

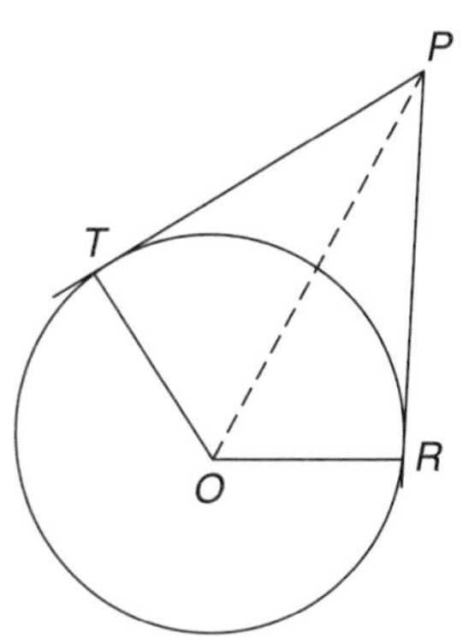

Figure 9.5

$$PT = PR$$

$$\overline{OT} \perp \overline{PT} \qquad \overline{OR} \perp \overline{PR}$$

Use Figure 9.5 and the given information to find the specified measurement.

1. If $PT = 12$, find PR.
2. If $PR = 28$, find PT.
3. If $PT = 12$ and $OT = 5$, find PO.
4. If $PO = 35$ and $PR = 28$, find OR.
5. If $m\angle TOP = 53°$, find $m\angle TRP$.
6. If $m\angle TOP = 45°$ and $TO = 8$, find TP.
7. If $PT = 3x + 5$ and $PR = 73 - x$, find x.
8. If $PT = 70 - 2x$ and $PR = \frac{1}{2}(x - 5)$, find PR.
9. If $PT = x + 10$, $OP = 2x - 7$, and $OT = x + 1$, find x.
10. If $PR = 2x - 16$, $OP = 2x - 9$, and $OR = x - 15$, find PT.

Two chords

When two chords intersect in a circle, the chords divide each other in such a way that the product of the lengths of the segments of one chord is equal to the product of the lengths of the segments of the other chord. If you look at Figure 9.6 and draw in $\overline{AC}$ and $\overline{BD}$, you can prove that $\angle CAB$ and $\angle CDB$ are congruent because they are both equal to one-half the measure of $\overset{\frown}{CB}$, and $\angle AEC \cong \angle DEB$ because they're vertical angles. That means $\Delta AEC \sim \Delta DEB$ by AA, and as a result,

$$\frac{AE}{ED} = \frac{CE}{EB}$$

Cross-multiplying gives you $AE \cdot EB = CE \cdot ED$.

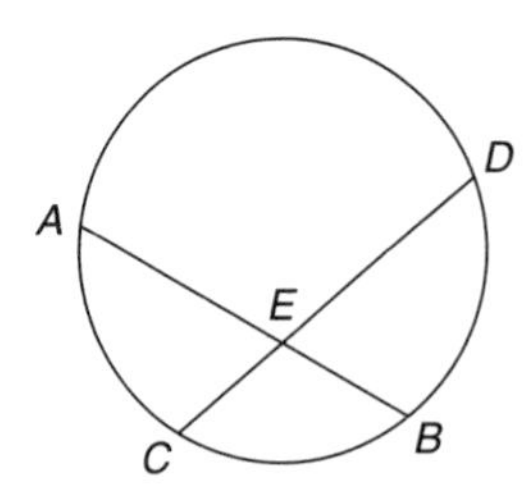

Figure 9.6

A diameter perpendicular to a chord bisects the chord and its arc. In Figure 9.7, if you know that diameter $\overline{VW}$ is perpendicular to chord $\overline{XY}$ and you draw in radii $\overline{OX}$ and $\overline{OY}$, you can prove $\Delta OZX \cong \Delta OZY$ by HL. By CPCTC, $\overline{XZ} \cong \overline{ZY}$ and $\angle XOZ \cong \angle YOZ$. Since the angles are congruent, their arcs are congruent, so $\overset{\frown}{XW} \cong \overset{\frown}{WY}$.

The converse is also true. If two chords intersect in such a way that one is the perpendicular bisector of the other, then the perpendicular bisector is a diameter.

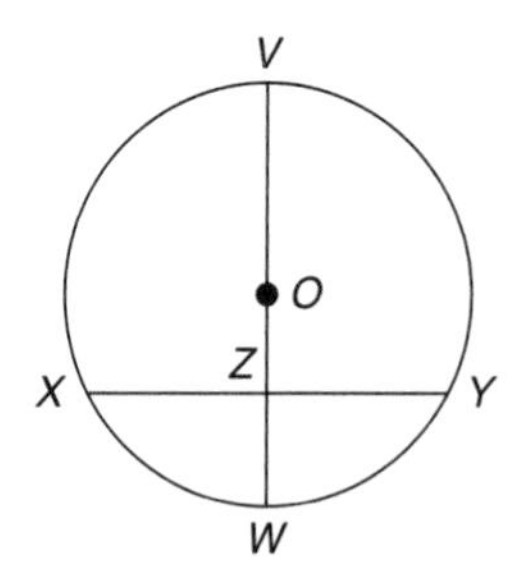

Figure 9.7

Use Figure 9.6 and the given information to find the specified measurement.

1. If $AE = 8$, $EB = 6$, and $CE = 12$, find ED.
2. If $AE = 9$, $ED = 6$, and $CE = 12$, find EB.
3. If $CD = 16$, $EB = 3$, and $AE = 13$, find CE.
4. If $AE = 5x - 11$, $EB = 2x + 1$, $CE = 6x - 2$, and $ED = x + 2$, find x.
5. If $AE = 10x$, $EB = 2x$, $CE = 7x - 1$, and $ED = 4x$, find x.

Use Figure 9.7 and the given information to find the specified measurement.

6. If $XZ = ZY = 10$ and $VZ = 25$, find VW.
7. If $XZ = ZY = 3$ and $OV = 5$, find ZW.
8. If $VZ = 8$ and $ZW = 18$, find XY.
9. If $XZ = x - 1$, $VZ = 2x - 2$, and $ZW = x - 8$, find x.
10. If $ZY = x - 3$, $VZ = x - 9$, and $ZW = x + 6$, find x.

Two secants

A secant, by definition, is a line, and lines have infinite length, so what we're really talking about here is a secant segment, or an extended chord. It has an internal segment, the chord, and an external segment, which is the portion outside the circle, from the point where the secants intersect to the first time it cuts the circle. When two secants are drawn to a circle from a single point outside the circle, the product of the length of the external segment of the secant and the full length of the secant is the same for each secant.

A secant and a tangent

When a secant and a tangent are drawn to a circle from a single point, the product of the lengths of the external segment of the secant and the entire secant is equal to the square of the length of the tangent segment. The tangent doesn't have an internal segment because it just touches the circle, so its external segment and its whole length are the same thing.

Use the following diagram and the given information to find the specified measurement.

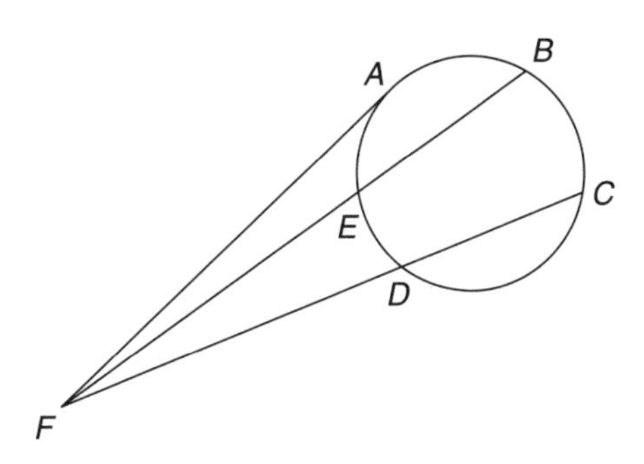

1. If $FE = 14$, $FB = 21$, and $FD = 7$, find FC.
2. If $FE = 11$, $EB = 4$, and $CD = 28$, find FD.
3. If $FA = 12$ and $FD = 8$, find DC.
4. If $FE = 9$ and $EB = 40$, find FA.
5. If $FB = 13$, $FD = 4$, and $DC = 22$, find FE.
6. If $FE = 3$, $FB = x + 7$, $DC = x$, and $FD = 4$, find FC.
7. If $FE = 2x$, $EB = 3x - 6$, $FD = 2x + 5$, and $CD = x + 11$, find FC.
8. If $FA = 6x - 2$, $FC = 7x + 1$, and $FD = 5x - 3$, find DC.
9. If $FE = x - 3$, $EB = x - 1$, and $FA = x + 1$, find FA.
10. If $FE = x$, $FB = 7x - 2$, $FD = x + 4$, and $DC = 3x$, find FE.

Equation of a circle

Because a circle is the set of all points at a fixed distance from a center point, using the distance formula with the center (h, k) and a point on the circle (x, y) produces the equation that describes a circle with center (h, k) and radius r:

$$(x-h)^2+(y-k)^2=r^2$$

Find the center and radius of each circle.

1. $x^2+y^2=4$
2. $x^2+y^2=16$
3. $(x-2)^2+(y-5)^2=9$
4. $(x-7)^2+(y+1)^2=25$
5. $(x+3)^2+(y+9)^2=1$

Write the equation of the circle with the given center and radius.

6. Center (2, 7), radius = 3
7. Center (–3, 4), radius = 2.5
8. Center (9, –7), radius = 8
9. Center (–6, –4), radius = $\sqrt{13}$
10. Center (0, 2), radius = $7\sqrt{3}$

Trigonometry

·10·

The word *trigonometry* means triangle measurement. For any acute angle, you can show that all right triangles containing an acute angle of that size are similar, and as a result, the ratios of different pairs of sides will be same. The ratio of the side opposite the acute angle to the hypotenuse, for example, will be the same for all triangles with a particular acute angle. Figure 10.1 shows that the hypotenuse is always opposite the right angle, but the labels "opposite" and "adjacent" depend upon the angle you use.

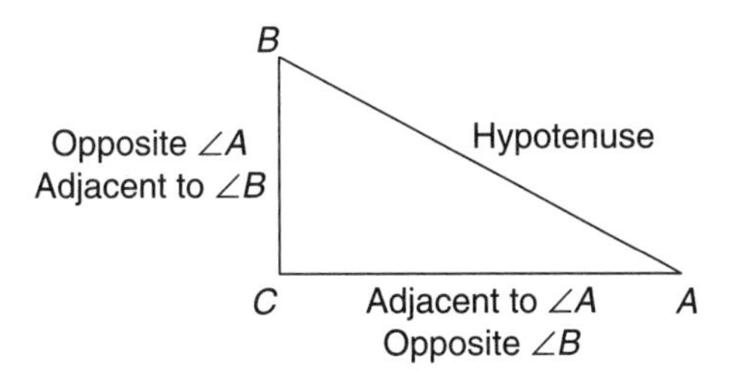

Figure 10.1

Ratios

For each acute angle in the right triangle, there are six ratios possible, but three of them—sine, cosine, and tangent—are used more often than the others:

$$\sin A = \frac{\text{opposite leg}}{\text{hypotenuse}} = \frac{BC}{AB} \qquad \sin B = \frac{\text{opposite leg}}{\text{hypotenuse}} = \frac{AC}{AB}$$

$$\cos A = \frac{\text{adjacent leg}}{\text{hypotenuse}} = \frac{AC}{AB} \qquad \cos B = \frac{\text{adjacent leg}}{\text{hypotenuse}} = \frac{BC}{AB}$$

$$\tan A = \frac{\text{opposite leg}}{\text{adjacent leg}} = \frac{BC}{AC} \qquad \tan B = \frac{\text{opposite leg}}{\text{adjacent leg}} = \frac{AC}{BC}$$

If the measurement of the acute angle is known, the value of any one of the ratios can be found from a table or by the use of a scientific calculator. If the lengths of the sides are known, you can work backward to find out the measurement of the angle.

EXERCISE

10·1

In ΔXYZ, $\angle Y$ is a right angle, XY = 3 cm, and YZ = 4 cm. Find the value of each ratio.

1. $\sin Z$
2. $\cos X$
3. $\tan X$
4. $\cos Z$
5. $\sin X$

In ΔRST, $\angle T$ is a right angle, RT = 5 in, and ST = 12 in. Find the value of each ratio.

6. $\tan R$
7. $\sin S$
8. $\cos R$
9. $\tan S$
10. $\sin R$

Find the value of each ratio in a table or by using the sin, cos, or tan keys on a scientific calculator.

11. $\sin 37°$
12. $\tan 45°$
13. $\cos 60°$
14. $\sin 15°$
15. $\tan 83°$

Finding a side of a right triangle

If you know the measurement of an acute angle and the measurement of one side of a right triangle, you can use trigonometric ratios to find the lengths of the other sides. Choose the ratio formed by the side you know and the side you need to find. Write the definition of the ratio, fill in the known side and the value of the ratio (from a table or your calculator), and solve for the missing side.

If you have a right triangle that contains a 30° angle and you know the hypotenuse is 12 in long, but you want to find the side opposite the 30° angle, you can use the sine ratio to find that length. (Of course, you could also realize it's a 30°-60°-90° right triangle and use the relationships you memorized, but pretend you didn't notice that.)

$$\sin 30° = \frac{\text{opposite}}{\text{hypotenuse}}$$

$$0.5 = \frac{x}{12}$$

$$x = 6$$

In △ABC, ∠C is a right angle. Use the given information to find the missing side.

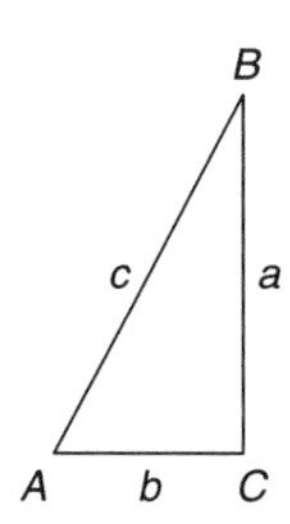

1. Find a if $c = 18$ and $\angle A = 84°$.
2. Find b if $a = 132$ and $\angle A = 18°$.
3. Find a if $b = 96$ and $\angle B = 35°$.
4. Find b if $c = 73$ and $\angle B = 46°$.
5. Find a if $c = 103$ and $\angle B = 28°$.
6. Find c if $a = 82$ and $\angle A = 75°$.
7. Find a if $b = 78$ and $\angle A = 33°$.
8. Find c if $a = 36$ and $\angle B = 59°$.
9. Find b if $a = 39$ and $\angle B = 42°$.
10. Find c if $b = 40$ and $\angle A = 24°$.
11. A ladder 24 ft long leans against a building, making an angle of 74° with the ground. How high up on the wall does the ladder reach?
12. A kite string attached to the ground makes an angle of 61° with the ground. When the full 500 ft of string has been played out, how high is the kite?
13. The diagonal of a rectangle makes a 64° angle with the side of the rectangle that measures 39 in. Find the length of the diagonal.
14. A hill has a slope of 12°. The distance up the hillside from the foot of the hill to the top of the hill is 3800 ft. How high is the hill?
15. Two radii in a circle meet at an angle of 54°. Find the length of the chord that joins the ends of the radii on the circle if the radius measures 10 cm. (*Hint*: A diameter perpendicular to a chord bisects the chord and its arc.)

Finding an angle of a right triangle

If you know the lengths of two sides of a right triangle, you can use trigonometric ratios to find the measurement of the acute angles. Choose the ratio that uses the two sides you know, and fill in the known sides to find the value of the ratio. If you use tables, divide the side lengths and carry the division to four decimal digits. Then look through the appropriate column of your table to find the decimal closest to your ratio, and trace across to find the angle. If you use a calculator, you can find the measurement of the angle by using the $\sin^{-1}$, $\cos^{-1}$, or $\tan^{-1}$ keys. If $\sin A = \frac{3}{5}$, then $A = \sin^{-1}\left(\frac{3}{5}\right)$.

EXERCISE

10·3

In ΔRST, ∠T is a right angle. Use the given information to find the specified angle.

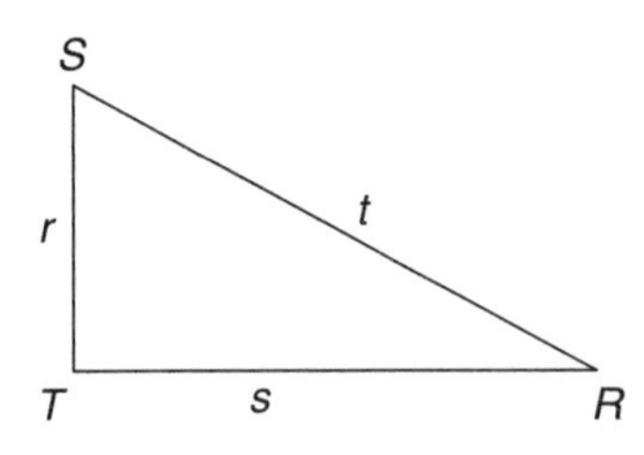

1. If $r = 3$ cm and $t = 6$ cm, find $\angle R$.
2. If $s = 4$ in and $t = 7$ in, find $\angle R$.
3. If $r = 8$ cm and $s = 6$ cm, find $\angle R$.
4. If $s = 5$ in and $r = 2$ in, find $\angle S$.
5. If $s = 12$ in and $t = 15$ in, find $\angle S$.
6. If $t = 18$ cm and $r = 6$ cm, find $\angle R$.
7. If $t = 32$ in and $s = 8$ in, find $\angle R$.
8. If $t = 325$ m and $r = 120$ m, find $\angle S$.
9. If $t = 90$ mi and $s = 45$ mi, find $\angle S$.
10. If $r = 10$ cm and $t = 14$ cm, find $\angle R$.
11. A rectangle is 65 cm long and 27 cm wide. What angle does the diagonal make with the longer side?
12. A ladder 20 ft long reaches to a point 16 ft up a wall. What angle does the ladder make with the ground?
13. To prevent accidents, a safety sticker advises that the foot of a ladder be placed 1 ft from the wall for every 4 ft of the ladder's length. What angle will the ladder make with the ground?
14. Find the larger acute angle of a right triangle whose legs are 27 and 55 in.
15. Find the measure of one of the base angles of an isosceles triangle whose legs are each 72 in and whose base is 48 in. (*Hint*: Draw the altitude to the base.)

Nonright triangles

All the work in this chapter so far was based on the assumption that you were working with right triangles. It is possible, however, to use trigonometry to solve for unknown sides or angles in nonright triangles. Two other rules allow you to solve such problems under the right conditions.

Law of sines

In Chapter 5, when you studied inequalities in triangles, you learned that the largest angle of a triangle was opposite the longest side, and the smallest angle opposite the shortest side. The law of sines takes this idea a step further. It says that the ratio of a side to the sine of the opposite angle is constant throughout the triangle.

$$\frac{a}{\sin(A)} = \frac{b}{\sin(B)} = \frac{c}{\sin(C)}$$

If you have two sides and the angle opposite one of them or if you know two angles and the side opposite one of them, as in Figure 10.2, you may be able to use a proportion to find the missing measurements.

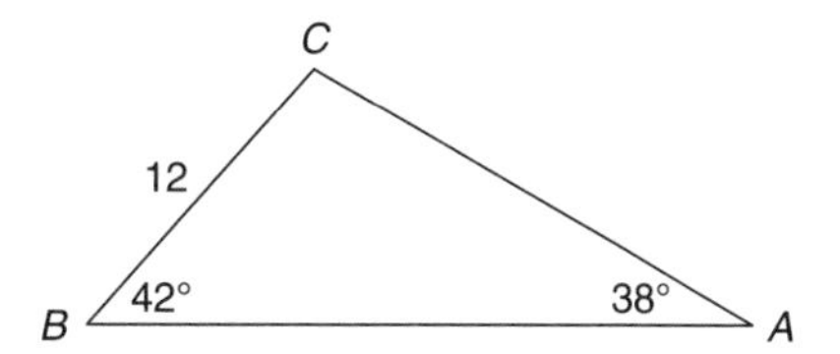

Figure 10.2 Two angles and a side not included between them

In ΔABC, $m\angle A = 38°$, $m\angle B = 42°$, and $BC = 12$ cm. Find the length of $\overline{AC}$.

$\overline{BC}$ is the side opposite $\angle A$, and you must find the length of $\overline{AC}$, the side opposite $\angle B$. Use the proportion $a/\sin(A) = b/\sin(B)$ with $a = 12$, $m\angle A = 38°$, and $m\angle B = 42°$, $a = 12$, $m\angle A = 38°$, and $m\angle B = 42°$. Then $12/\sin(38°) = b/\sin(42°)$. Cross-multiply to solve the proportion.

$$12\sin(42°) = b\sin(38°)$$

$$\frac{12\sin(42°)}{\sin(38°)} = b$$

$$\frac{12(0.6691)}{0.6157} \approx b$$

$$\frac{8.0292}{0.6157} \approx b$$

$$13.041 \approx b$$

You can also use the law of sines to find a missing angle. In ΔXYZ, $XY = 15$ cm, $YZ = 22$ cm, and $XZ = 9$ cm. If $\angle Y$ measures 18°, you can find the measures of the other angles of the triangle by the law of sines.

$$\frac{YZ}{\sin(X)} = \frac{XZ}{\sin(Y)} = \frac{XY}{\sin(Z)}$$

$$\frac{22}{\sin(X)} = \frac{9}{\sin(18°)} = \frac{15}{\sin(Z)}$$

Work first with $9/\sin(18^\circ) = 15/\sin(Z)$, and $\sin(Z) = [15\sin(18^\circ)]/9 \approx 0.5150$. Use the inverse function on your calculator to find that $\sin^{-1}(0.5150) \approx 31^\circ$. Once you know the measures of two angles, $\angle X$ is $180^\circ - (18^\circ + 31^\circ) = 180^\circ - 49^\circ = 131^\circ$.

The ambiguous case

The law of sines is useful when you know the sizes of two sides and one angle or two angles and one side. You have to take care to examine all the possibilties, however. The results of the law of sines can be ambiguous if the given information is two sides and an angle other than the included angle. If you were given ΔABC with $m\angle A = 30^\circ$, $AB = 12$, and $BC = 8$, you could do the calculation.

$$\frac{BC}{\sin(A)} = \frac{AB}{\sin(C)}$$

$$\frac{8}{\sin(30^\circ)} = \frac{12}{\sin(C)}$$

$$12\sin(30^\circ) = 8\sin(C)$$

$$\frac{12\sin(30^\circ)}{8} = \sin(C)$$

$$\frac{12(0.5)}{8} = \sin(C)$$

$$0.75 = \sin(C)$$

Your calculator will tell you that $\sin^{-1}(0.75) = 48.6^\circ$. It is possible that $\angle C = 48.6^\circ$, but it is also possible that $\angle C = 180 - 48.6^\circ = 131.4^\circ$. Both angles—$48.6^\circ$ and 131.4°—have a sine equal to 0.75. There are actually two triangles that could fit the initial description. Figure 10.3 shows two different triangles that both fit the specifications.

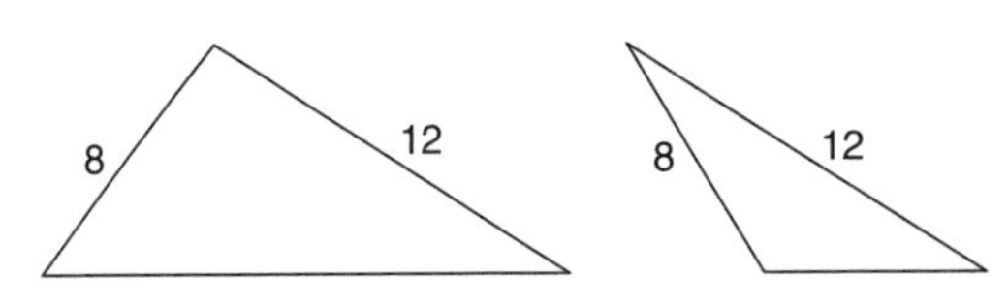

Figure 10.3

Whenever you use the law of sines to solve for a missing angle, always stop to consider whether the supplement of the angle you've found is also a possible angle of the triangle. Sometimes there is only one triangle possible with the given measurements, but sometimes there are two distinct triangles that can be created with those measurements. Remember that $\sin(X) = \sin(180 - X)$ and watch for the other possible triangle.

Do you remember the shortcuts for proving triangles congruent? The law of sines will give a unique solution for the triangle when the given information is SAS, ASA, or AAS. Those are all sufficient conditions for congruent triangles. The ambiguous case occurs when you have SSA, which is not a way of proving triangles congruent.

You also need to guard against trying to do the impossible. If in ΔABC, $AB = 15$ ft, $BC = 22$ ft, and $\angle C$ measures 50°, applying the law of sines will give you $15/\sin 50^\circ = 22/\sin A$ and $\sin A = (22\sin 50^\circ)/15 \approx 1.1235$. Now that's a problem, since the values of the sine never go above 1. You have to conclude that it's impossible to have a triangle with these measurements.

Law of cosines

The law of sines is a powerful tool, and the calculations involved are not too complicated, but it does require that you know at least one angle. What if you know all three sides, but none of the angles, and you want to find the angles? Turn to the law of cosines.

The law of cosines has a familiar look. It resembles the right-triangle relationship, the Pythagorean theorem, but with an extra term. The law of cosines says that in a triangle with sides a, b, and c opposite $\angle A$, $\angle B$, and $\angle C$, respectively,

$$c^2 = a^2 + b^2 - 2ab\cos(C)$$

You could learn three versions of that formula. There's $a^2 = b^2 + c^2 - 2bc\cos(A)$ and $b^2 = a^2 + c^2 - 2ac\cos(B)$ as well as the one above. If you just remember that you want to start with a side and end with the angle opposite that side, you can modify the formula easily enough.

The law of cosines is most useful when you know the lengths of all three sides and need to find an angle, or when you know two sides and the included angle and want to find the third side. If ΔXYZ has sides of length 15, 22, and 35 and you want to find the measure of the largest angle of the triangle, remember that the largest angle will be opposite the longest side, which measures 35. Start with the 35 as c, because you want to end with $\angle C$.

$$\begin{aligned} c^2 &= a^2 + b^2 - 2ab\cos(C) \\ 35^2 &= 15^2 + 22^2 - 2\cdot 15\cdot 22\cdot \cos(C) \\ 1{,}225 &= 225 + 484 - 660\cos(C) \\ 1{,}225 &= 709 - 660\cos(C) \end{aligned}$$

The most common error in law of cosine problems is a violation of the order of operations. Many people, when they reach the line $1{,}225 = 709 - 660\cos(C)$, will mistakenly subtract 660 from 709. Don't be one of them. The multiplication of 660 cos(C) would have to be done before any addition or subtraction. Instead, subtract 709 from both sides, then divide by –660 to isolate cos(C).

$$\begin{aligned} 1{,}225 &= 709 - 660\cos(C) \\ 1{,}225 - 709 &= -660\cos(C) \\ 516 &= -660\cos(C) \\ \frac{-516}{660} &= \cos(C) \\ -0.7818 &= \cos(C) \end{aligned}$$

Use the inverse cosine key on your calculator to find the measure of $\angle C$.

$$\begin{aligned} C &= \cos^{-1}(-0.7818) \\ C &\approx 141.4^\circ \end{aligned}$$

To use the law of cosines to find a side, you'll need to know the angle opposite that side. In ΔABC, if $a = 4.2$ cm, $b = 6.3$ cm, and $\angle C$ measures 115°, you can find the length of side c by substituting into the law of cosines.

$$c^2 = a^2 + b^2 - 2ab\cos(C)$$
$$c^2 = (4.25)^2 + (6.25)^2 - 2(4.25)(6.25)\cos(116^\circ)$$
$$c^2 = 18.0625 + 39.0625 - 53.1250\cos(116^\circ)$$
$$c^2 = 57.1250 - 53.1250(-0.4388)$$
$$c^2 = 57.1250 + 23.3113$$
$$c^2 = 80.4363$$
$$c = \sqrt{80.4363} \approx 8.9686$$

The third side measures just a bit less than 9 cm.

Use the following figure to answer questions 1 and 2.

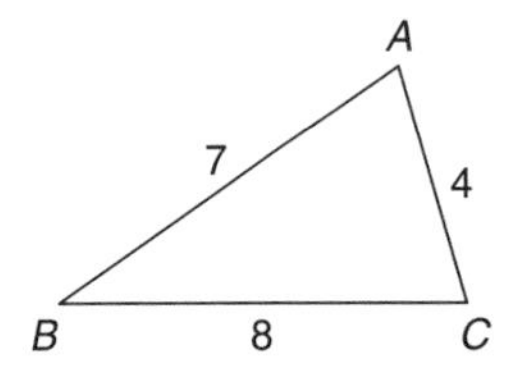

1. Find the measure of $\angle A$.
2. Find the measure of $\angle B$.

Use the following figure to answer questions 3 and 4.

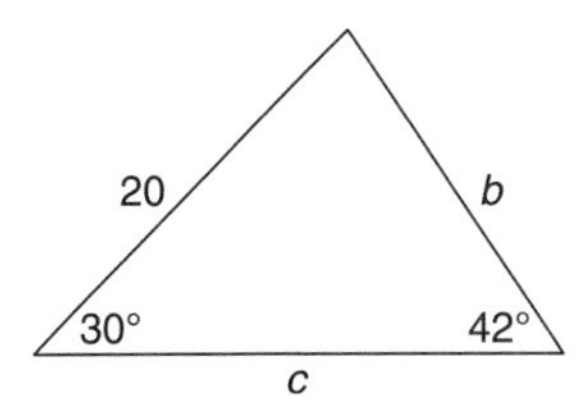

3. Find the length of side b.
4. Find the length of side c.
5. In ΔRST, side s measures 6 in, side r measures 9 in, and $\angle T$ is 165°. Find the length of side t.
6. In ΔABC, $\angle A$ measures 65° and $\angle C$ is 10°. If $BC = 13$ cm, find the lengths of the other two sides.
7. If ΔXYZ has a 30° angle at vertex Y, $XZ = 8$ cm, and $YZ = 16$ cm, find all possible measurements for the angles of the triangle.
8. If ΔXYZ is an obtuse triangle with $XY = 42$, $YZ = 53$, and $m\angle Y = 19°$, find the length of $\overline{XZ}$.

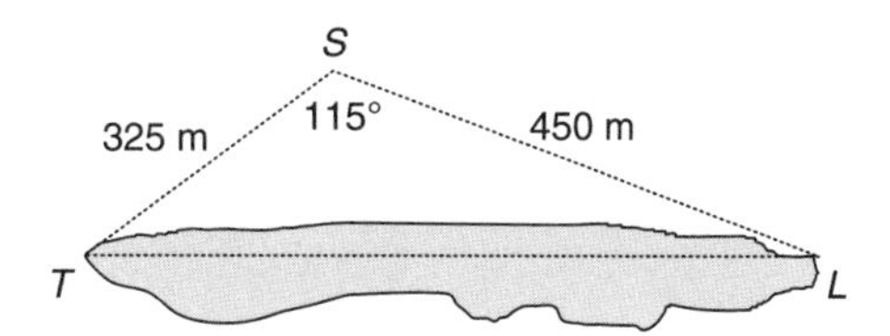

9. From point S on one side of a lake, the distance to either end of the lake is measured (see the figure above). If $ST = 325$ m, $SL = 450$ m, and $\angle S$ measures 115°, find the width of the lake, from L to T, to the nearest meter.

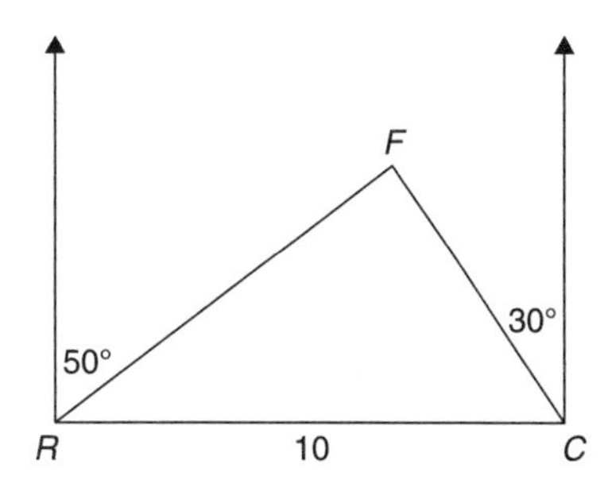

Ranger station *R*, campsite *C*, and fire at *F*.

10. In the figure above, a forest ranger spots a fire 50° east of north. A group of campers sees the same fire from their campsite, 30° west of north. If the campers are 10 miles due east of the ranger's station, find the ranger's distance from the fire.

Area of triangles and parallelograms

To find the area of a triangle or a parallelogram, you need to know the length of a base and the length of the height drawn perpendicular to that base. Unfortunately, you don't always have that information. If you know the lengths of two sides and the angle between them, however, and you remember your trigonometry, you can still find the area.

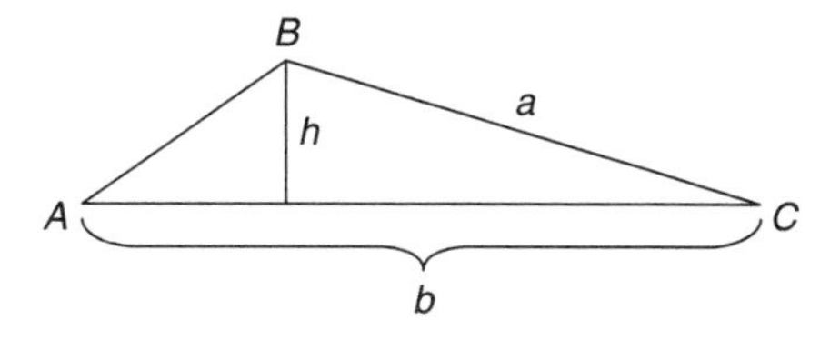

Figure 10.4 Triangle with height *h*

When you draw the altitude, or height, from a vertex of the triangle perpendicular to the opposite side, as in Figure 10.4, you form a right triangle. Right-triangle trigonometry tells you that $\sin(C) = h/a$ so $h = a\sin(C)$. The area of the triangle is $\frac{1}{2}bh = \frac{1}{2}b \cdot a\sin(C)$ or $A = \frac{1}{2}ab\sin(C)$, where a and b are adjacent sides and $\angle C$ is the angle between them.

You can easily show that the area of a parallelogram is $A = ab\sin(C)$, where a and b are adjacent sides and $\angle C$ is the angle between them.

EXERCISE

10·5

Solve each problem. Include units in your answer.

1. Find the area of ΔABC if $AB = 8$ cm, $BC = 12$ cm, and $\angle B$ measures 30°.
2. Find the area of a parallelogram whose adjacent sides measure 4 in and 7 in, if the angle between them measures 48°.
3. Find the area of a rhombus with sides of 14 cm, if the smaller angle of the rhombus measures 56°.
4. Find the area of ΔRST if $ST = 18$ cm, $RT = 24$ cm, and $\angle T$ measures 15°.
5. Find the area of $\square PQRS$ if $PQ = 36$ in, $QR = 15$ in, and $\angle Q$ measures 125°.
6. Find the length of side $\overline{BC}$ in ΔABC if $AB = 4$ cm, $\angle B$ measures 30°, and the area of ΔABC is 17 cm^2.
7. If a rhombus has sides of length 18 in and an area of 296 square inches(in^2), find the measure of the acute angle of the rhombus to the nearest degree.
8. If the area of an isosceles triangle is 49 cm^2 and the vertex angle measures 130°, find the lengths of the congruent legs.
9. Find the area of an equilateral triangle with a side of 9 in.
10. If the area of a parallelogram with angles of 70° and 110° is 751.75 cm^2, and the longer side measures 40 cm, find the length of the shorter side to the nearest centimeter.

Coordinate geometry

·11·

The Cartesian coordinate system locates every point in the plane by an ordered pair of numbers (x, y) in which the x coordinate indicates horizontal movement and the y coordinate vertical movement. You used the Cartesian coordinate system to graph equations in algebra, but you can also explore geometry on the coordinate plane.

Distance

The distance between two points can be calculated by the distance formula $d=\sqrt{(x_2-x_1)^2+(y_2-y_1)^2}$. The formula is an application of the Pythagorean theorem, in which the difference of the x coordinates gives the length of one leg of a right triangle and the difference of the y coordinates the length of the other. The distance between (x_1, y_1) and (x_2, y_2) is the hypotenuse of the right triangle. If the two points fall on a vertical line or on a horizontal line, the distance will simply be the difference in the coordinates that don't match.

The distance between the points (4, –1) and (0, 2) is

$$d=\sqrt{(4-0)^2+(-1-2)^2}=\sqrt{16+9}=5$$

Find the distance between the given points.

1. (4, 5) and (7, –4)
2. (6, 2) and (7, 6)
3. (–7, –1) and (–5, –6)
4. (5, 3) and (8, –2)
5. (–4, 2) and (3, 2)

Given the distance between the two points, find the possible values for the missing coordinate.

6. $(a, -2)$ and (7, 2) are 5 units apart.
7. (–1, 3) and $(4, d)$ are 13 units apart.
8. (8, –6) and $(c, -6)$ are 7 units apart.
9. $(2, b)$ and (2, –1) are 9 units apart.
10. (a, a) and (0, 0) are $4\sqrt{2}$ units apart.

Midpoint

The midpoint of the segment that connects (x_1, y_1) and (x_2, y_2) can be found by averaging the x coordinates and averaging the y coordinates.

$$M = \left(\frac{x_1 + x_2}{2}, \frac{y_1 + y_2}{2} \right)$$

The midpoint of the segment connecting (4, –1) and (0, 2) is

$$M = \left(\frac{4+0}{2}, \frac{-1+2}{2} \right) = \left(2, \frac{1}{2} \right)$$

EXERCISE
11·2

Find the midpoint of the segment with the given endpoints.

1. (2, 3) and (5, 8)
2. (–3, 1) and (–1, 8)
3. (–5, –3) and (–1, –1)
4. (8, 0) and (0, 8)
5. (0, –2) and (4, –4)

Given the midpoint M of the segment connecting A and B, find the missing coordinate.

6. *A*(*x*, 6), *B*(6, 8), *M*(4, 7)
7. *A*(–1, 3), *B*(*x*, 9), *M*(3, 6)
8. *A*(–5, *y*), *B*(7, –3), *M*(1, 3)
9. *A*(4, –9), *B*(–2, *y*), *M*(1, –7)
10. *A*(0, 4), *B*(*x*, 0), *M*(8, 2)

Slope

The *slope m* of a line is a measurement of the rate at which it rises or falls. A rising line has a positive slope while a falling line has a negative slope. The larger the absolute value of the slope, the steeper the line. A horizontal line has a slope of zero, and a vertical line has an undefined slope.

$$m = \frac{\text{rise}}{\text{run}} = \frac{y_2 - y_1}{x_2 - x_1}$$

The slope of the line through the points (4, –1) and (0, 2) is

$$m = \frac{2-(-1)}{0-4} = -\frac{3}{4}$$

EXERCISE
11·3

Find the slope of the line that passes through the two points given.

1. (–5, 5) and (5, –1)
2. (6, –4) and (9, –6)
3. (3, 4) and (8, 4)
4. (4, 6) and (8, 7)
5. (7, 2) and (7, 5)

If a line has the given slope and passes through the given points, find the missing coordinate.

6. $m = -4$, $(4, y)$ and $(3, 2)$
7. $m = 2$, $(2, 9)$ and $(x, 13)$
8. $m = \frac{1}{2}$, $(-4, 1)$ and $(3, y)$
9. $m = -\frac{3}{5}$, $(x, 0)$ and $(2, -6)$
10. $m = 0$, $(-4, y)$ and $(-6, 3)$

Equation of a line

Linear equations are equations in two variables whose graphs are lines. There are three forms for the linear equation: slope-intercept form, standard form, and point-slope form.

Slope-intercept form

Slope-intercept form, or $y = mx + b$ form, is probably the most common form of the equation of a line. It is useful for graphing, because the slope and y intercept of the graph can be read directly from the equation. In the $y = mx + b$ form, m is the slope and b is the y intercept. In the equation $y = \frac{2}{3}x - 4$, you can read that the slope of the line is $\frac{2}{3}$ and its y intercept is $(0, -4)$. Linear equations that are not in slope-intercept form can be transformed to slope-intercept form by solving for y.

If you have to graph the equation $y = \frac{2}{3}x - 4$, you can begin by placing a point at the y intercept of $(0, -4)$ and then counting out the slope, moving up 2 units and 3 units right. Place a point there and repeat (see Figure 11.1, left). Once you have a few points, use a straightedge to connect them and extend the line. Place arrows on the ends to indicate that the line continues (see Figure 11.1, right).

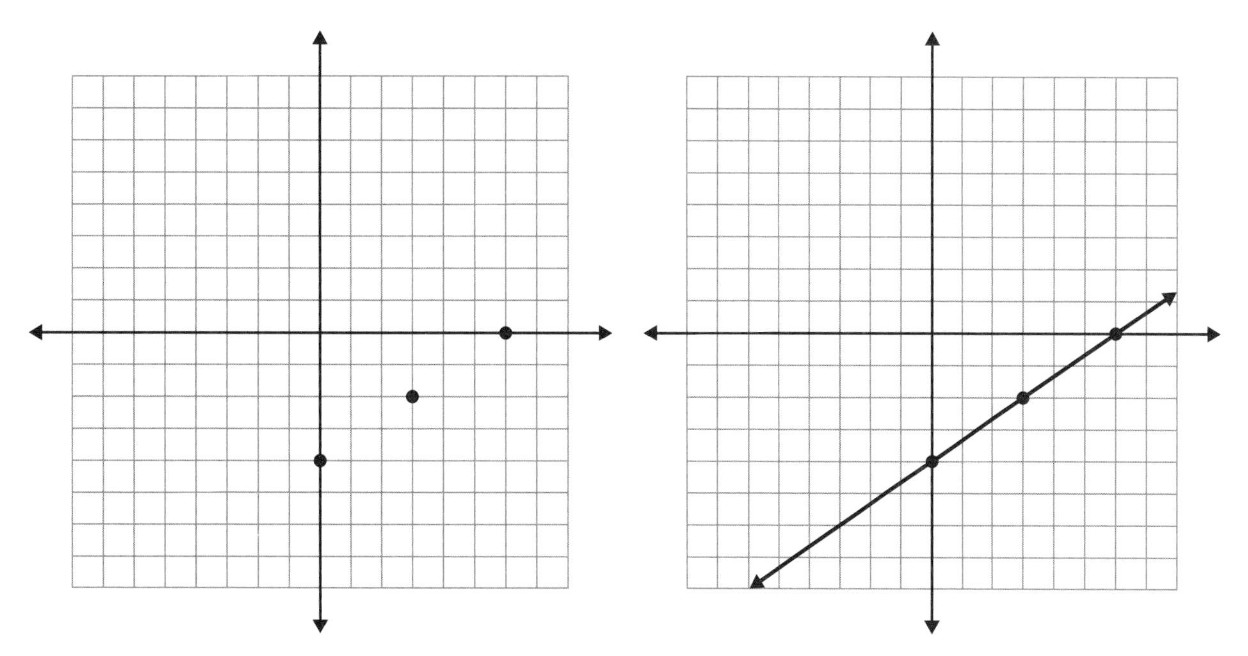

Figure 11.1 Start from the *y* intercept, plot points by counting the slope, and then draw the line.

Standard form

A linear equation is in *standard form* when the x and y terms are on the same side of the equation, equal to a constant, and all the coefficients are integers. In symbols, standard form is $ax + by = c$, where a, b, and c are integers, so the equation $4x - 9y = 18$ is in standard form, but $5y - 3 = 2x$ is not.

The quickest way to graph a linear equation in standard form is to find and plot the x and y intercepts. The x intercept is the point on the graph where $y = 0$, and the y intercept is the point where $x = 0$. To find the x intercept of $4x - 9y = 18$, replace y with 0, and solve. So $4x - 9(0) = 18$ becomes $4x = 18$ and $x = 4.5$. To find the y intercept, replace x with 0 and solve. Now $4(0) - 9y = 18$ becomes $-9y = 18$ so $y = -2$. Plot both the x and y intercepts, connect, and extend to graph the line, as in Figure 11.2.

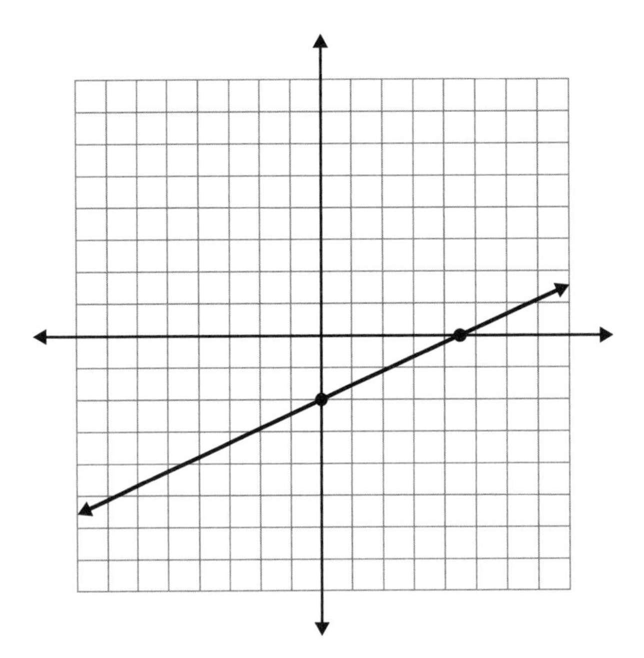

Figure 11.2 Plot the x and y intercepts and connect them with a line.

EXERCISE

11·4

Use slope and the y intercept to graph each linear equation.

1. $y = 2x - 3$
2. $y = -3x + 7$
3. $y = \frac{3}{4}x - 5$
4. $y = -\frac{1}{2}x + 4$
5. $y = 5 - x$

Use x and y intercepts to graph equations in standard form.

6. $4x + 3y = 12$
7. $5x - 2y = 10$
8. $x + 6y = 3$
9. $3x - 5y = -15$
10. $2x - 3y = 9$

Point-slope form

The third form of the linear equation is not particularly useful for graphing, but it's the best form to use when you are asked to write the equation of a line whose graph is shown to you, or when you're asked for the equation of a line that has a certain slope or passes through certain points.

The *point-slope form* of a line with a slope m that passes through the point (x_1, y_1) is $y - y_1 = m(x - x_1)$. Replace m with the given slope, x_1 with the x value of the given point, and y_1 with the y value of the given point. Once you've done that, technically you've got the equation of the line. Most of the time, you'll want to simplify, however, and put that equation in either slope-intercept or standard form.

To find the equation of the line with slope $-\frac{1}{2}$ that passes through the point (−4, 3), begin with point-slope form and plug in the given values.

$$y - y_1 = m(x - x_1)$$
$$y - 3 = -\tfrac{1}{2}(x + 4)$$

Simplify by distributing to remove the parentheses, and work toward the form you want to end up with.

$$y - 3 = -\tfrac{1}{2}x - 2$$

If you want slope-intercept form, add 3 to both sides: $y = -\frac{1}{2}x + 1$. If you'd prefer standard form, get the x and y terms on the same side and the constant term on the other: $\frac{1}{2}x + y = 1$. Then multiply the entire equation by 2, to eliminate the fraction: $x + 2y = 2$.

If you are given two points rather than the slope and a point, use the slope formula to find the slope of the line connecting the two points. Then choose either one of the given points to use in the point-slope form with the slope you found. To find the equation of the line passing through (−4, −3) and (5, 15), first calculate the slope

$$m = \frac{y_2 - y_1}{x_2 - x_1} = \frac{15 + 3}{5 + 4} = \frac{18}{9} = 2$$

Then use that slope and either point in the point-slope form. If you choose (−4, −3), you get $y + 3 = 2(x + 4)$, and if you choose (5, 15), you get $y - 15 = 2(x - 5)$. Although they look different in this form, when you simplify, both become $y = 2x + 5$.

EXERCISE

11·5

1. Write the equation in standard form of the line with a slope of −3 through the point (−2, 7).
2. Write the equation in slope-intercept form of the line with a slope of $\frac{1}{3}$ through the point (6, 4).
3. Write the equation in standard form of the line with a slope of $-\frac{2}{5}$ through the point (5, −1).
4. Write the equation in slope-intercept form of the line with a slope of $\frac{5}{2}$ through the point (−4, −5).
5. Write the equation in standard form of the line through the points (−3, −2) and (2, 8).
6. Write the equation in slope-intercept form of the line through the points (−4, 3) and (2, −3).
7. Write the equation in standard form of the line through the points (4, 7) and (−5, −2).
8. Write the equation in slope-intercept form of the line through the points (−8, 3) and (7, −2).
9. Write the equation in standard form of the line through the points (−7, 9) and (5, 0).
10. Write the equation in slope-intercept form of the line through the points (−1, 10) and (3, −8).

Parallel and perpendicular lines

Parallel lines have the same slope. Perpendicular lines have slopes that multiply to –1, that is, slopes that are negative reciprocals. To find the equation of a line parallel to or perpendicular to a given line, first determine the slope of the given line. Be sure the equation is in slope-intercept form before you try to determine the slope. Use the same slope for a parallel line or the negative reciprocal for a perpendicular line, along with the given point, in the point-slope form of a line $y - y_1 = m(x - x_1)$.

To find a line parallel to $y = 3x - 7$ that passes through the point (4, –1), use the slope of 3 from $y = 3x - 7$ and the point (4, –1) in point-slope form and simplify.

$$\begin{aligned} y-(-1) &= 3(x-4) \\ y+1 &= 3x-12 \\ y &= 3x-13 \end{aligned}$$

To find a line perpendicular to $y = 3x - 7$ that passes through the point (4, –1), use a slope of $-\frac{1}{3}$.

$$\begin{aligned} y-(-1) &= -\frac{1}{3}(x-4) \\ y+1 &= -\frac{1}{3}x+\frac{4}{3} \\ y &= -\frac{1}{3}x+\frac{1}{3} \end{aligned}$$

Determine whether the lines are parallel, perpendicular, or neither.

1. $y = \frac{1}{3}x - 2$ and $3x + y = 7$
2. $x - 5y = 3$ and $2x - 10y = 9$
3. $2y - 8x = 9$ and $4y = 3 - 18x$
4. $4y = 6x - 7$ and $y = \frac{3}{2}x + 5$
5. $y = \frac{4}{5}x + 8$ and $5x + 4y = 8$

Find the equation of the line described.

6. Parallel to $y = 5x - 3$ and passing through the point (3, –1)
7. Perpendicular to $6y - 8x = 15$ and passing through the point (–4, 5)
8. Parallel to $4x + 3y = 21$ and passing through the point (1, 1)
9. Perpendicular to $y = 4x - 3$ and passing through the point (4, 13)
10. Parallel to $2y = 4x + 16$ and passing through the point (8, 0)

Coordinate proof

The formulas of coordinate geometry can be used to prove conjectures. Using the distance formula to find the lengths of two segments, you can prove that they are congruent. You can prove lines are parallel by showing they have the same slope, or that they are perpendicular by showing their slopes are negative reciprocals. To show that two segments bisect each other, show they have the same midpoint.

Coordinate proofs are generally done without specific number coordinates, so that the results will be generally true and not just specific to one set of points. The techniques are the same, however; you'll just be working with letters instead of numbers. Create the initial diagram carefully to make your work as simple as possible.

1. Prove that the diagonals of a parallelogram bisect each other. [Place the parallelogram with one vertex at the origin and one side on the positive x axis. Label the vertices $(0, 0)$, $(a, 0)$, (b, c), and $(a + b, c)$.]

2. Prove that if the altitude of a triangle bisects the side to which it is drawn, the triangle is isosceles. [Place the triangle with one side on the x axis and the altitude on the y axis. Vertices of the triangle are $(-a, 0)$, $(a, 0)$, and $(0, b)$.]

3. Prove that the diagonals of a rectangle are congruent. [Vertices of the rectangle are $(0, 0)$, $(a, 0)$, $(0, b)$, and (a, b).]

4. Prove that the line segment joining the midpoints of two sides of a triangle is parallel to the third side and one-half as long. [Vertices of the triangle are $(0, 0)$, $(0, a)$, and (b, c). Find the midpoints, slopes, and lengths.]

5. Prove that the diagonals of a rhombus are perpendicular. [Vertices of the rhombus are $(0, 0)$, $(c, 0)$, (a, b), and $(a + c, b)$, but because all sides are congruent, $c = \sqrt{a^2 + b^2}$.]

Transformations

·12·

The word *transformation* is a general term that includes several operations that move or change a polygon. These include *reflection*, *translation*, and *rotation*, which are rigid transformations that do not change the shape of the polygon, and dilation, which stretches or shrinks the polygon. Rigid transformations produce images that are congruent to the original, but dilations produce images that are the same shape but a different size than the original or preimage.

Reflection

A *reflection* mirrors points across a specified line called the *reflecting line*. The image of a point P is a point P' positioned on the other side of the reflecting line so that $\overline{PP'}$ is perpendicular to the reflecting line and the preimage P and the image P' are equidistant from the reflecting line. Imagine looking at your reflection in a mirror. The "other you" is directly across from you and appears to be as far into the mirror as you are in front of it. Reflection preserves length and angle measure, but reverses orientation, just as your mirror image seems to have left and right backward.

When you look at transformations on the coordinate plane, you can predict what will happen to the entire polygon by just looking at what happens to its vertices. Reflection over the x axis changes the sign of the y coordinate of each point, and reflection over the y axis changes the sign of the x coordinate. Reflection over the line $y = x$ swaps the x and y coordinates, and reflection over the line $y = -x$ swaps the coordinates and changes the signs of both coordinates.

Figure 12.1 shows the quadrilateral $A(-4, 2)$, $B(0, 3)$, $C(5, -1)$, and $D(-2, 0)$ and its image when reflected in the x axis, when reflected in the y axis, when reflected across the line $y = x$, and when reflected across the line $y = -x$.

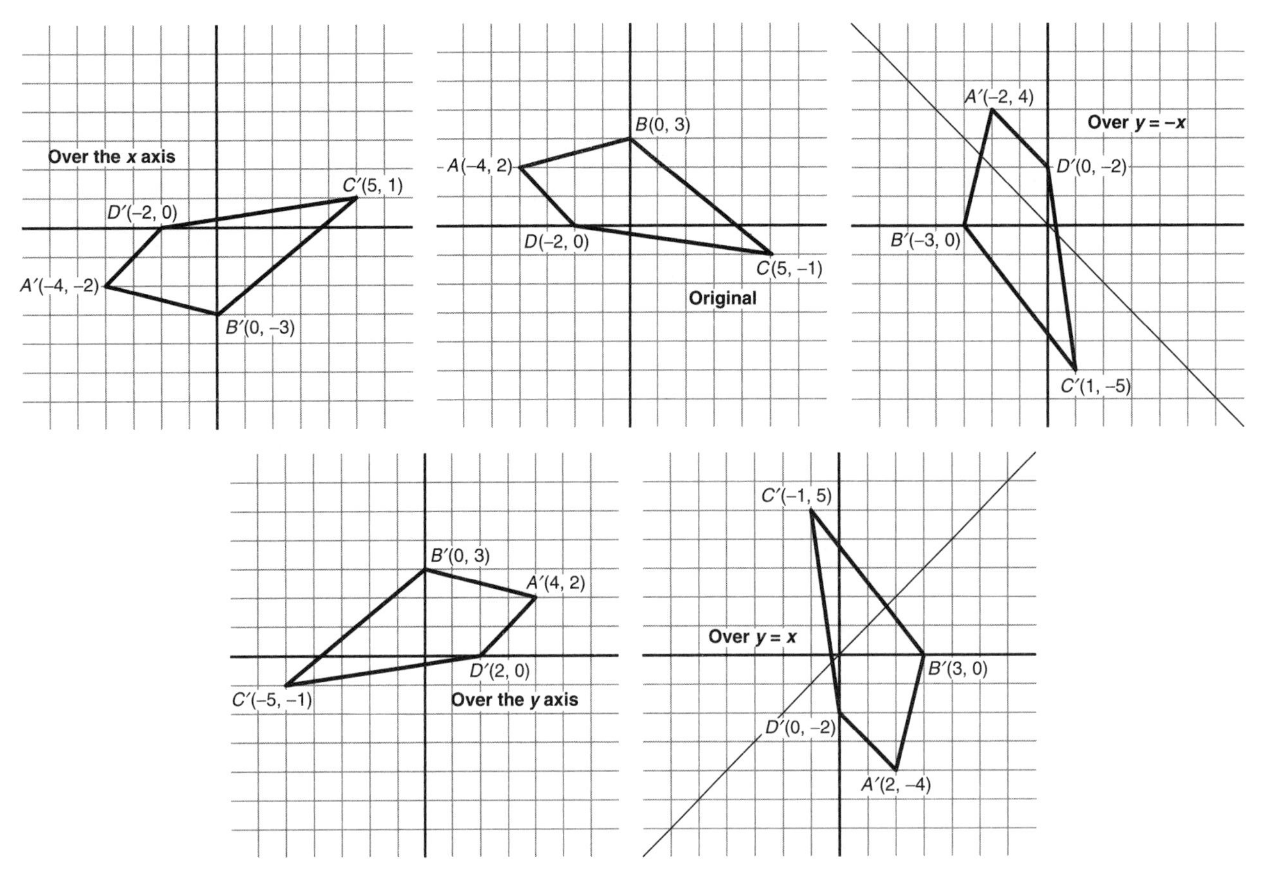

Figure 12.1 Quadrilateral *ABCD* and its reflection over different lines.

EXERCISE

12·1

Find the reflection of the given point over the given line.

1. $P(2, -3)$ over the x axis
2. $Q(-1, 6)$ over the y axis
3. $R(4, 3)$ over the line $y = x$
4. $S(5, -1)$ over the line $y = -x$
5. $T(0, -2)$ over the x axis.

The vertices of a polygon are given. Find the coordinates of the vertices of its image after a reflection over the specified line.

6. $A(2, -3)$, $B(-1, 7)$, $C(4, 3)$ over the y axis
7. $A(4, 5)$, $B(-3, 2)$, $C(2, -3)$ over the x axis
8. $A(1, 7)$, $B(-2, 5)$, $C(-3, -4)$, $D(3, -2)$ over the line $y = x$
9. $A(-1, 1)$, $B(7, 1)$, $C(7, -2)$, $D(-1, -2)$ over the x axis
10. $A(1, 4)$, $B(3, 1)$, $C(2, -4)$, $D(-2, -3)$, $E(-1, 3)$ over the y axis

Translation

Translation is actually the composition of two reflections across parallel reflecting lines. It preserves length and angle measure, and since the first reflection reverses orientation and the second reflection reverses it again, the orientation of the final image is the same as that of the original, or preimage.

Translation is generally visualized as a sliding or shifting and is denoted by indicating the horizontal and the vertical component of the movement. $T(x, y) \to (x + h, y + k)$ indicates that every point (x, y) is moved h units horizontally and k units vertically. If the h value is positive, the point moves right; and if h is negative, the point moves left. A positive value for k indicates a move up, and a negative value shows a downward move.

Find the image of the given point under the specified translation.

1. $P(3, 7)$ under $T(x, y) \to (x + 5, y - 2)$
2. $P(2, 7)$ under $T(x, y) \to (x - 3, y - 1)$
3. $P(-3, 1)$ under $T(x, y) \to (x + 2, y + 5)$
4. $P(-4, -3)$ under $T(x, y) \to (x - 4, y + 3)$
5. $P(2, 0)$ under $T(x, y) \to (x + 1, y - 4)$

Given a translation and the image point, find the preimage.

6. $T(x, y) \to (x + 5, y + 3)$, $P'(8, -3)$
7. $T(x, y) \to (x - 1, y + 4)$, $P'(2, 7)$
8. $T(x, y) \to (x + 3, y - 5)$, $P'(-4, -2)$
9. $T(x, y) \to (x - 2, y - 1)$, $P'(8, -7)$
10. $T(x, y) \to (x + 4, y - 2)$, $P'(-5, 9)$

The vertices of a polygon are given. Find the coordinates of the vertices of its image after the specified translation.

11. $A(2, -3)$, $B(-1, 7)$, $C(4, 3)$ under $T(x, y) \to (x - 1, y + 5)$
12. $A(4, 5)$, $B(-3, 2)$, $C(2, -3)$ under $T(x, y) \to (x + 3, y - 2)$
13. $A(1, 7)$, $B(-2, 5)$, $C(-3, -4)$, $D(3, -2)$ under $T(x, y) \to (x + 2, y + 3)$
14. $A(-1, 1)$, $B(7, 1)$, $C(7, -2)$, $D(-1, -2)$ under $T(x, y) \to (x - 5, y - 4)$
15. $A(1, 4)$, $B(3, 1)$, $C(2, -4)$, $D(-2, -3)$, $E(-1, 3)$ under $T(x, y) \to (x + 4, y - 1)$

Rotation

Rotation is the composition of two reflections across reflecting lines that intersect. The preimage is rotated around a point, in either the clockwise or the counterclockwise direction. The amount of rotation is measured in degrees, and the acute angle formed by the two reflecting lines is one-half the number of degrees of the rotation.

Rotations are specified by the center point around which the points are rotated and the number of degrees in the rotation. Rotation in the counterclockwise direction is considered positive, while clockwise rotation is considered negative.

R_{90}: A rotation of 90° about the origin moves the point from (x, y) to $(-y, x)$.
R_{180}: A rotation of 180° about the origin moves the point from (x, y) to $(-x, -y)$.

R_{270}: A rotation of 270° about the origin moves the point from (x, y) to $(y, -x)$.
R_{-90}: A rotation of −90° about the origin moves the point from (x, y) to $(y, -x)$.
R_{-180}: A rotation of −180° about the origin moves the point from (x, y) to $(-x, -y)$.
R_{-270}: A rotation of −270° about the origin moves the point from (x, y) to $(-y, x)$.

Notice that $R_{-270} = R_{90}$, $R_{270} = R_{-90}$, and $R_{-180} = R_{180}$, so you really need to learn only three. Figure 12.2 shows ΔABC rotated about the origin.

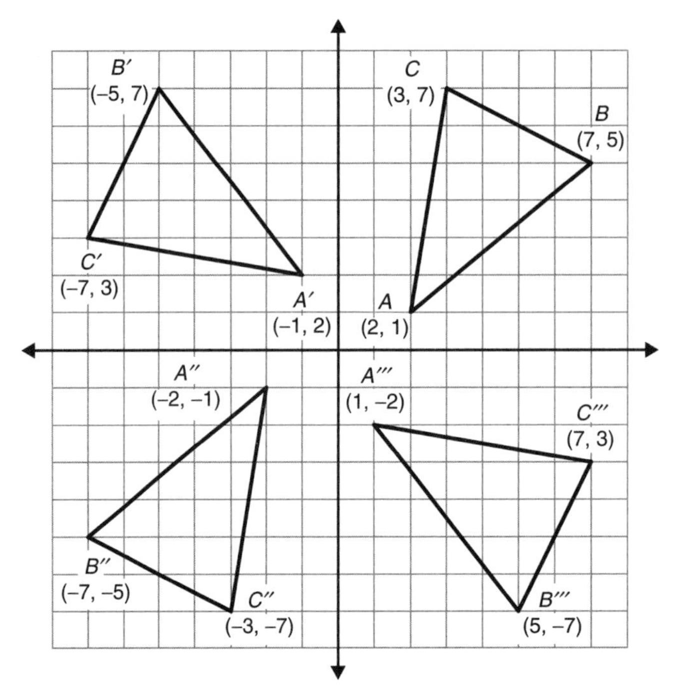

Figure 12.2

EXERCISE 12·3

Find the image of the given point under the specified rotation about the origin.

1. (4, –1) under R_{90}
2. (–3, 2) under R_{-90}
3. (1, 0) under R_{180}
4. (0, –4) under R_{270}
5. (–3, –1) under R_{-270}

Find the image of the polygon under the specified rotation about the origin.

6. $A(2, -3)$, $B(-1, 7)$, $C(4, 3)$ under R_{270}
7. $A(4, 5)$, $B(-3, 2)$, $C(2, -3)$ under R_{-90}
8. $A(1, 7)$, $B(-2, 5)$, $C(-3, -4)$, $D(3, -2)$ under R_{90}
9. $A(-1, 1)$, $B(7, 1)$, $C(7, -2)$, $D(-1, -2)$ under R_{-180}
10. $A(1, 4)$, $B(3, 1)$, $C(2, -4)$, $D(-2, -3)$ $E(-1, 3)$ under R_{-270}

Dilation

A *dilation* produces a scaled version of the preimage. If the scale factor is greater than 1, the image will be larger than the preimage. If the scale factor is between 0 and 1, the image will be smaller than the preimage.

To specify a dilation, give the center of the dilation and a scale factor. If rays are drawn from the center of dilation through the vertices of the polygon, the distance from the center to each vertex can be calculated. Each distance is multiplied by the scale factor to produce the distance from the center to the image, and then the image vertices are located on the rays.

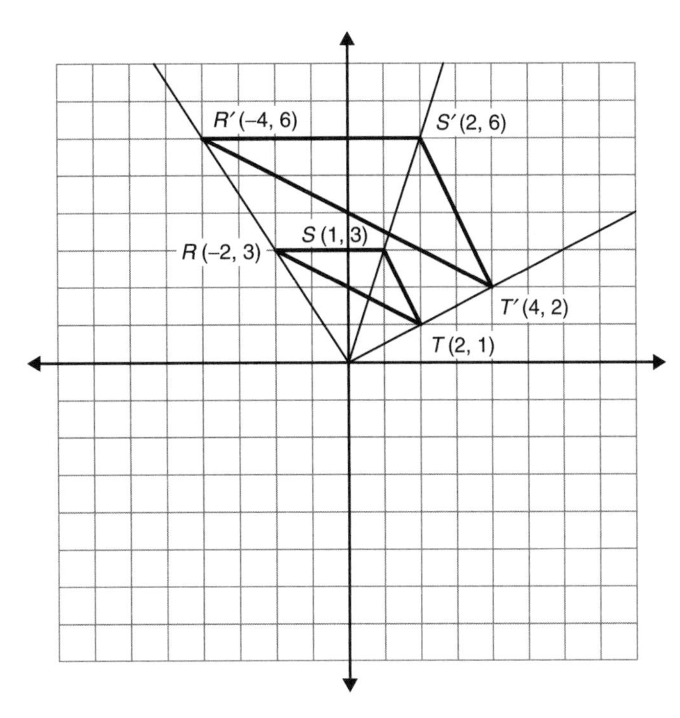

Figure 12.3 A dilation centered at the origin.

If the center is the origin, the effect of a dilation with a scale factor of k is to send the point (x, y) to (kx, ky). (See Figure 12.3.) If the center is the point (x_c, y_c) and the scale factor is k, the image of each point (x, y) will be $(x_c + k(x - x_c), y_c + k(y - y_c))$. Figure 12.4 shows a dilation centered at (2, –4).

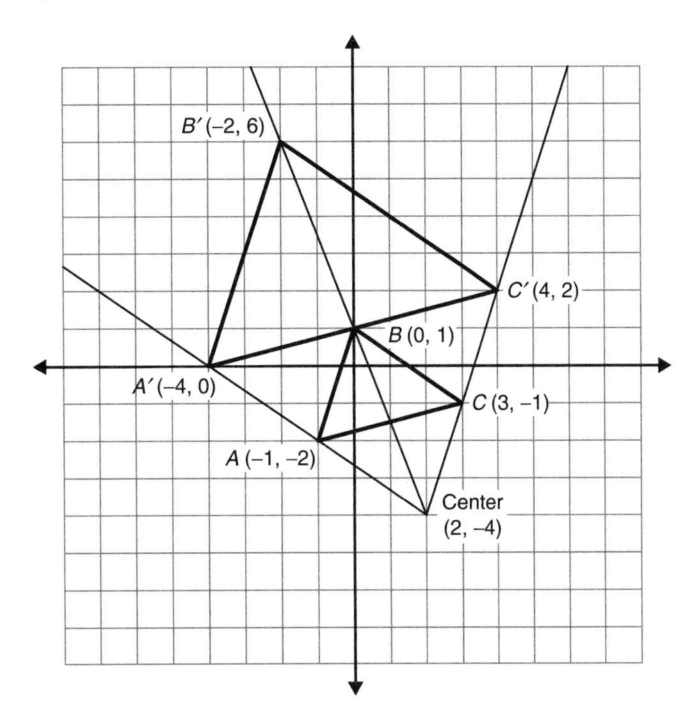

Figure 12.4 A dilation centered at (2, –4).

EXERCISE

12·4

Find the image of each polygon under the given dilation centered at the origin, where D_k *denotes a dilation by scale factor* k.

1. $A(2, -3)$, $B(-1, 7)$, $C(4, 3)$ under D_2
2. $A(4, 5)$, $B(-3, 2)$, $C(2, -3)$ under $D_{1/2}$
3. $A(1, 7)$, $B(-2, 5)$, $C(-3, -4)$, $D(3, -2)$ under $D_{1.5}$
4. $A(-1, 1)$, $B(7, 1)$, $C(7, -2)$, $D(-1, -2)$ under $D_{0.8}$
5. $A(1, 4)$, $B(3, 1)$, $C(2, -4)$, $D(-2, -3)$, $E(-1, 3)$ under D_3

Find the image of each polygon under the dilation with center and scale factor indicated.

6. $A(2, -3)$, $B(-1, 7)$, $C(4, 3)$, center at A, $k = 2$
7. $A(4, 5)$, $B(-3, 2)$, $C(2, -3)$, center at C, $k = 1.5$
8. $A(1, 7)$, $B(-2, 5)$, $C(-3, -4)$, $D(3, -2)$, center at D, $k = 1/2$
9. $A(-1, 1)$, $B(7, 1)$, $C(7, -2)$, $D(-1, -2)$, center at $(7, 0)$, $k = 1.1$
10. $A(1, 4)$, $B(3, 1)$, $C(2, -4)$, $D(-2, -3)$, $E(-1, 3)$, center at A, $k = 3$

Area, perimeter, and circumference

The perimeter of a polygon or the circumference of a circle measures the linear distance around the figure, while area measures the space enclosed by a figure.

Rectangles and squares

The area of a rectangle is equal to the product of its length and width, or base and height. Since a square is a rectangle with congruent sides, the product of the length and width becomes the length of any side squared. The perimeter of a rectangle is the sum of the lengths of the sides. Since the opposite sides are congruent, the perimeter is twice the length plus twice the width. The perimeter of a square is four times any side since all sides are congruent.

Rectangle: $A = lw = bh$ $\quad P = 2l + 2w$

Square: $A = s^2$ $\quad P = 4s$

Use area and perimeter formulas for rectangles and squares to solve each problem.

1. Find the area of a rectangle with a length of 14 cm and a width of 8 cm.
2. Find the area of a square with a side of 9 m.
3. Find the area of a rectangle with a length of 12 cm and a diagonal of 13 cm.
4. Find the area of a square with a diagonal of $6\sqrt{2}$ cm.
5. Find the width of a rectangle with a length of 13 cm and an area of 143 cm^2.
6. Find the side of a square with an area of 256 cm^2.
7. Find the diagonal of a square with an area of 128 cm^2.
8. The length of a rectangle is 7 cm more than its width. If the perimeter of the rectangle is 38 cm, find the area.

9. Each of the small squares has a perimeter of 16 cm, and the large square has a perimeter of 100 cm. Find the shaded area.

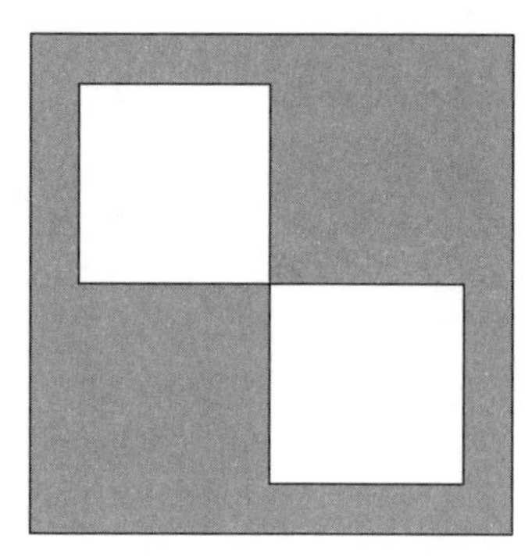

10. *ABCD* is a rectangle with perimeter of 34 cm. If the length is 2 more than twice the width, find the shaded area.

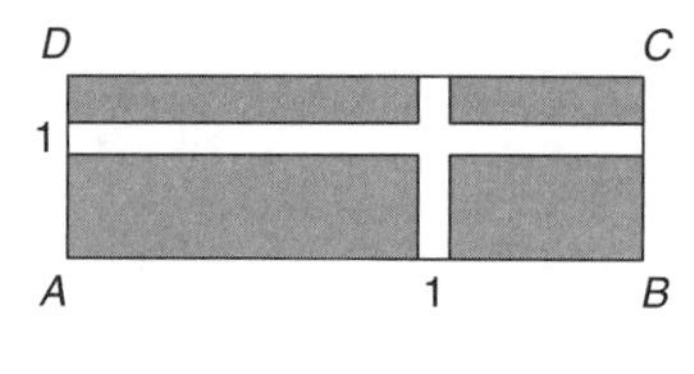

Parallelograms and trapezoids

In a parallelogram, any side may be the base. The height is the perpendicular distance from the base to its parallel partner. Do not confuse the height with the length of the adjacent side. In rectangles and squares you can use two sides as base and height because the sides meet at a right angle, but that's not true for parallelograms. The area of a parallelogram is the product of the lengths of a base and the altitude drawn perpendicular to that base. The perimeter is the sum of the lengths of the sides, and in the parallelogram, opposite sides are congruent; so if you denote the sides as a and b, the perimeter is twice a plus twice b. For a parallelogram, $A = bh$ and $P = 2a + ab$.

In a trapezoid, the height is the perpendicular distance between the parallel bases. The area of a trapezoid is one-half the product of the height and the sum of the bases. Since the length of the midsegment that connects the midpoints of the nonparallel sides is one-half the sum of the bases, the area can also be found by multiplying the length of the midsegment by the height. The perimeter of a trapezoid is the sum of the lengths of its sides, and there's no particular shortcut for that. For a trapezoid, $A = \frac{1}{2}h(b_1 + b_2)$.

Use the area formulas for parallelograms and trapezoids to solve each problem.

1. Find the area of a parallelogram with a base of 18 cm and a height of 12 cm.
2. Find the area of a trapezoid with bases of 3 and 7 m and a height of 5 m.

3. Find the area of a rhombus with a side of 14 cm and a height of 8 cm.
4. Find the area of a trapezoid with a longer base of 35 cm and a height and shorter base both equal to 19 cm.
5. Find the area of a parallelogram with sides of 12 and 25 cm and an included angle of 60°.
6. Find the height of a trapezoid with bases of 31 and 43 cm and an area of 740 cm^2.
7. Find the perimeter of a rectangle with a height of 9 cm and an area of 153 cm^2.
8. Find the length of the midsegment of a trapezoid with a height of 11 cm and an area of 209 cm^2.
9. *ABCD* is an isosceles trapezoid, with $\overline{CE} \parallel \overline{AD}$. The area of ▱*AECD* is 265 cm^2, and $EB = 6$ cm. If the area of the trapezoid is 340 cm^2, find the height.

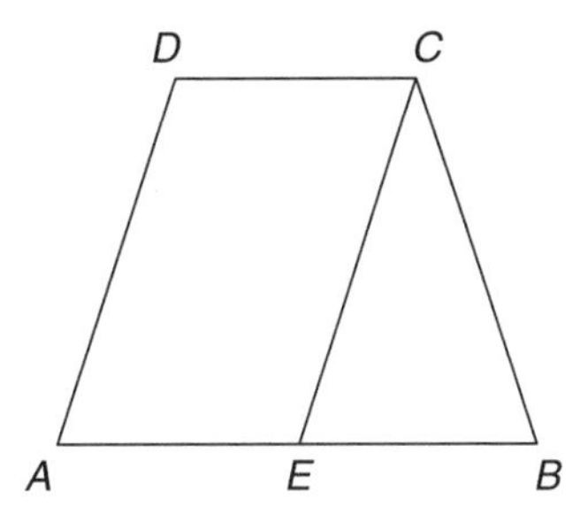

10. *WXYZ* is a parallelogram, and the white region is also a parallelogram. Find *ZY* if the shaded area is 294 cm^2, the height is 14 cm, and $RT = 2$ cm.

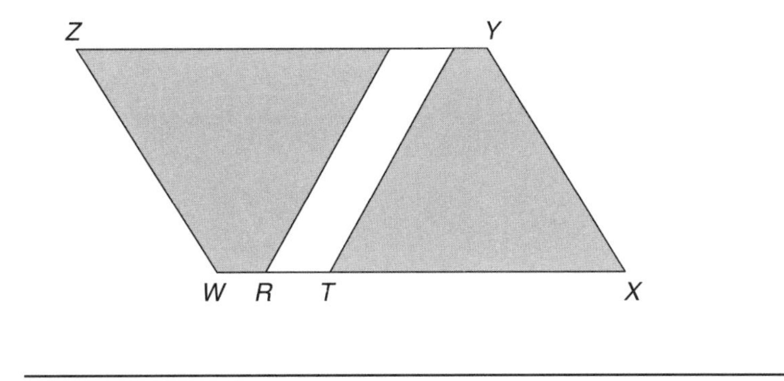

Triangles

The altitude of a triangle is a perpendicular from a vertex to the opposite side. It is sometimes necessary to extend a side to meet an altitude that falls outside the triangle.

The area of a triangle is one-half the product of the lengths of a base and the altitude drawn to that side. The perimeter of a triangle is the sum of the lengths of its sides. No shortcut exists for that rule either. For a triangle: $A = \frac{1}{2}bh$ and $P = a + b + c$.

The area of a triangle is one-half the area of a parallelogram with the same base and height. Drawing a diagonal in a parallelogram divides it into two congruent triangles, each with one-half the area of the parallelogram.

EXERCISE

13·3

Use the area formula for a triangle to solve each problem.

1. Find the area of a triangle with a base of 5 m and a height of 3 cm.
2. Find the area of a triangle with a base of 83 cm and a height of 42 cm.
3. Find the area of a right triangle with an acute angle of 30° if the shorter leg measures 8 cm.
4. Find the area of an isosceles triangle with a base of 42 cm and congruent legs each measuring 35 cm.
5. Find the area of an equilateral triangle with sides 16 cm long.
6. If a triangle with a base of 23 cm has an area of 149.5 cm^2, find the height.
7. If the area of a right triangle is 294 in^2 and one leg is 7 units longer than the other, find the length of the hypotenuse.
8. The area of ΔABC is 3 times the area of ΔABD. If the height of ΔABD is 7 m, find the height of ΔABC.

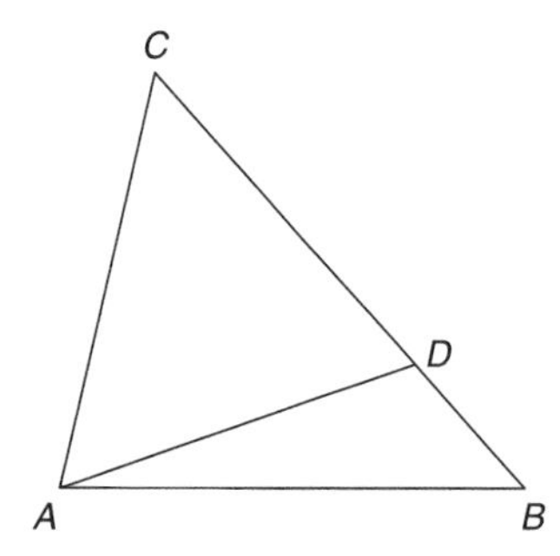

9. ΔXWY is isosceles with $XW = YW$. If $ZW = 3$ cm, $WV = 5$ cm, and $XY = 12$ cm, find the shaded area.

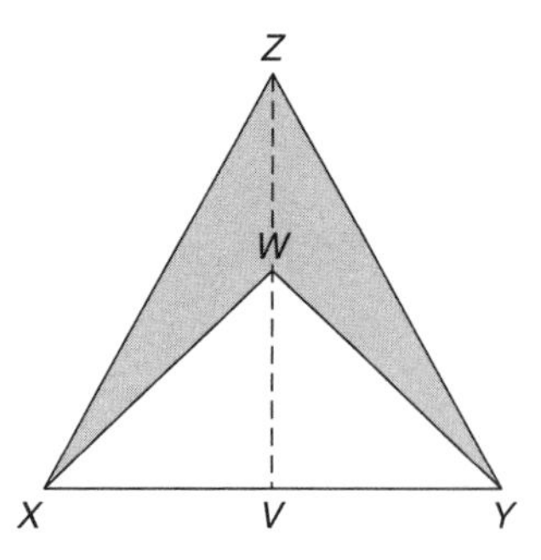

10. The base and height of a triangle are equal. If the area of the triangle is 48 cm^2, find the length of the base.

Regular polygons

A polygon is regular if all its sides are congruent and all its angles are congruent, if it is both equilateral and equiangular. The center of a regular polygon is a point in the interior of the polygon equidistant from all vertices. If you circumscribed the polygon, that is, if you drew a circle that passed through all the vertices of the regular polygon, the center of that circle would be the center of the polygon. The *radius* of a regular polygon is the distance from the center to a vertex, so the radius of the polygon is the radius of the circumscribed circle. The *apothem* is a segment from the center perpendicular to a side of the polygon.

The radius, the apothem, and one-half of a side of the polygon form a right triangle. If you know the measurements of two of the three, you can use the Pythagorean theorem to find the missing one. If you only know one of the three, you can calculate the measures of the acute angles in the right triangle and then use trigonometry to find missing measurements. The angle formed by the radius and the apothem has its vertex at the center of the regular polygon, and its measure is $360/2n = 180/n$, where n is the number of sides of the polygon.

To find the area of a regular polygon, you'll need to know the length of a side and the length of the apothem. The area of a regular polygon is equal to one-half the product of the length of the apothem and the perimeter. For a regular polygon, $A = \frac{1}{2}aP$.

Use the formula for the area of a regular polygon to solve each problem.

1. Find the area of a regular hexagon with a perimeter of 84 cm and an apothem of $7\sqrt{3}$ cm.
2. Find the area of a regular octagon with a side of 10 cm and an apothem of 12 cm.
3. Find the area of a regular pentagon with a radius of 13 cm and a side of 10 cm.
4. Find the area of a regular decagon with an apothem of 19 cm and a radius of 20 cm.
5. Find the area of an equilateral triangle with a radius of 4 cm and a side of $4\sqrt{3}$ cm.
6. Find the apothem of a regular hexagon with an area of $1{,}458\sqrt{3}$ cm^2 and a perimeter of $108\sqrt{3}$ cm.
7. Find, to the nearest centimeter, the apothem of a regular octagon with an area of 2028 cm^2 and a side of 21 cm.
8. Find the perimeter of a regular pentagon with an area of 292.5 cm^2 and an apothem of 9 cm.
9. Find the side of a regular decagon with an area of 2,520 cm^2 and an apothem of 28 cm.
10. Find the side of an equilateral triangle with an area of $36\sqrt{3}$ cm^2 and an apothem of $2\sqrt{3}$ cm.

Circles

The area of a circle is the product of the constant π and the square of the radius. The circumference, or linear distance around the circle, is the product of π and the diameter, or two times the product of π and the radius. If necessary, the value of π can be approximated as 3.14 or $\frac{22}{7}$, but since exact answers are preferred to approximations, you should leave answers in terms of π unless there's a compelling reason to use an approximation.

$$\text{Area: } A = \pi r^2$$

$$\text{Circumference: } C = \pi d = 2\pi r$$

A *sector* is a portion of a circle cut off by a central angle. For most people, the easiest image is a slice of cake or pie or pizza. An *arc* is a portion of the circumference of a circle, again cut off by a central angle. The measure of the central angle and the measure of the arc it intercepts are the same, and both are measured in degrees. The length of the arc depends on the size of the circle as well as the angle.

The area of a sector of a circle is a fraction of the area of the entire circle. The fraction is determined by the central angle of the sector. The length of an arc is a fraction of the circumference, also determined by the central angle.

$$\text{Area of a sector: } A = \frac{\text{angle measure}}{360} \cdot \pi r^2$$

$$\text{Length of an arc: } S = \frac{\text{angle measure}}{360} \cdot \pi d$$

Use the formulas for area, circumference, area of a sector, or length of an arc to solve each problem.

1. Find the area of a circle with radius of 5 cm.
2. Find the area of a circle with diameter of 22 m.
3. Find the area of a circle with circumference of 18π cm.
4. Find the radius of a circle with area of 49π m^2.
5. Find the diameter of a circle with area of 121π cm^2.
6. Find the circumference of a circle with area of 162π in^2.
7. A circular garden is to be covered with mulch. Each bag of mulch covers 4 ft^2 and costs \$1.85. If the garden has a diameter of 24 ft, what is the cost to mulch the garden?
8. If the circles in the figure below are concentric with radii as shown, find the shaded area.

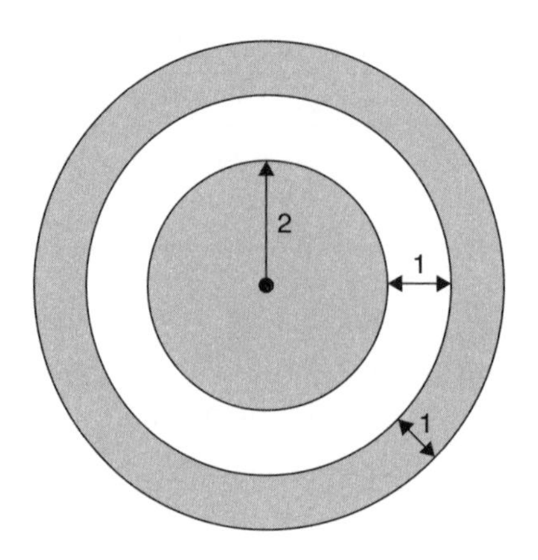

9. Find the shaded area in the following figure.

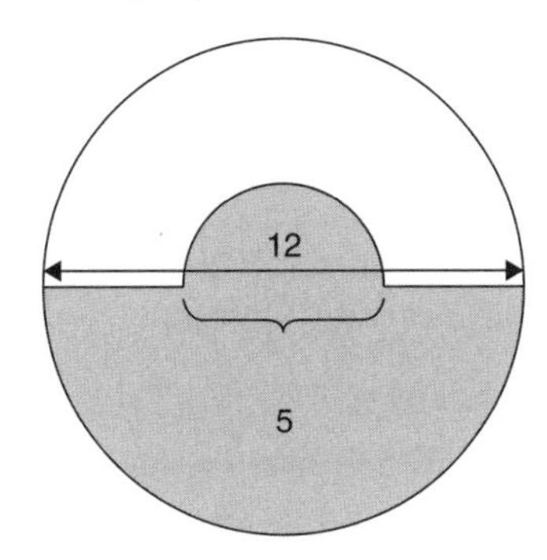

10. Find the area of a sector with a central angle of 40° and a radius of 18 cm.
11. Find the radius of a circle if a sector with a central angle of 120° has an area of 27π cm².
12. If a circle has a diameter of 28 cm, find the length of the arc intercepted by a central angle of 18°.
13. A central angle of 150° intercepts an arc 20π units long. Find the radius of the circle.
14. Find the circumference of a circle if a sector with a central angle of 45° has an area of 18π cm².
15. Find the area of a circle if a central angle of 10° intercepts an arc 12π cm long.

Areas and perimeters of similar polygons

If two polygons are similar, their corresponding sides are in proportion, that is, the ratio of their corresponding sides is constant. That ratio is called the *scale factor.* The perimeters of the similar polygons—the sums of the sides—have the same ratio. The areas of similar polygons, however, are proportional to the squares of the sides, because the area is the product of two linear measurements, such as length and width, or base and height. Each of those linear measurements gets multiplied by the scale factor. If the area of the smaller figure is bh and the scale factor is k, then the area of the larger figure is $kb \cdot kh = k^2bh$.

$$\text{Perimeters of similar figures: } \frac{\text{perimeter of polygon 1}}{\text{perimeter of polygon 2}} = \frac{\text{side of polygon 1}}{\text{side of polygon 2}}$$

$$\text{Areas of similar figures: } \frac{\text{area of polygon 1}}{\text{area of polygon 2}} = \frac{(\text{side of polygon 1})^2}{(\text{side of polygon 2})^2}$$

Use ratios to find the specified measurement.

1. The sides of two similar pentagons are in ratio 4:7. If the area of the smaller pentagon is 140 m², find the area of the larger pentagon.
2. The perimeters of two similar hexagons are in ratio 3:5. If the area of the larger hexagon is 365 cm², find the area of the smaller hexagon.

3. The areas of two similar decagons are in ratio 16:25. If the side of the smaller decagon is 28 in, find the side of the larger.
4. The areas of two similar octagons are in ratio 4:49. If the perimeter of the larger octagon is 952 m, find the perimeter of the smaller.
5. The areas of two similar quadrilaterals are in ratio 25:81. If the perimeter of the smaller quadrilateral is 95 ft, find the perimeter of the larger.
6. The sides of two similar triangles are in ratio 6:11. If the area of the larger triangle is 605 m^2, find the area of the smaller.
7. The sides of two similar pentagons are in ratio 8:13. If the area of the smaller pentagon is 448 ft^2, find the area of the larger.
8. The perimeter of a parallelogram is 74 cm, and its height is 11.5 cm. The sides of the parallelogram are enlarged proportionally until the perimeter is 222 cm. If the area of the larger parallelogram is 2380.5 cm^2, find the base of the smaller.
9. The perimeters of the two similar regular hexagons are in ratio 13:24. If the area of the smaller hexagon is 1183 cm^2, find the shaded area.

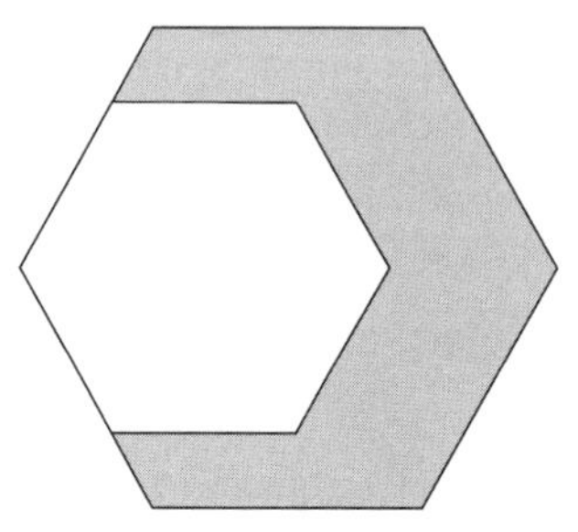

10. The sides of an octagonal window and the wooden frame surrounding it are in ratio 16:25. If the apothem of the window is 8 in and its area is 212 in^2, find the width of the frame, to the nearest 0.1 in.

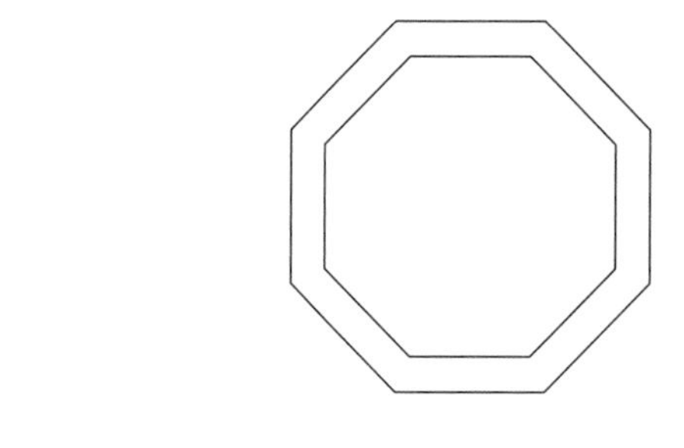

Areas of shaded regions

Problems that ask you to find the area of a shaded region are frequently used to ask you to apply several area formulas in combination. Most often these can be found by calculating the area of the overall figure and the area of the unshaded region and subtracting, but there are cases in which calculating the shaded area directly is the easier strategy. The circles in Figure 13.1 are congruent to one another, and tangent to one another and to the rectangle. If the diameter of each circle is 4 in, find the shaded area.

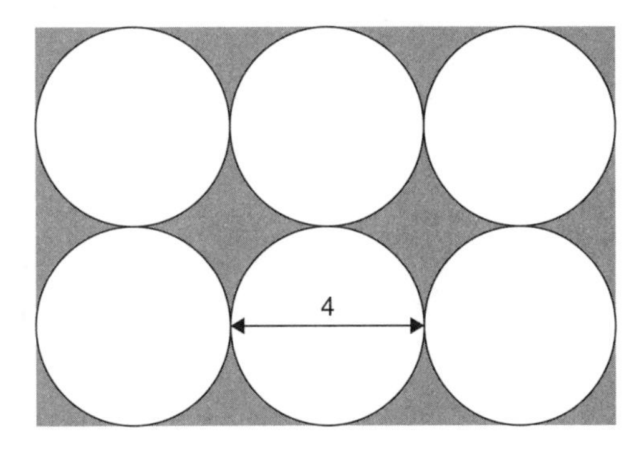

Figure 13.1

The area of the shaded region is the area of the rectangle minus the total area of the six circles. Each circle has a diameter of 4 and therefore a radius of 2 in. The area of one circle is 4π in^2, and the total area of the six circles is 24π in^2. The rectangle is three diameters long and two diameters wide, or 12 by 8 in. The area of the rectangle is 96 in^2. The shaded area is $96 - 24\pi$ or approximately 20.6 in^2.

1. In the following diagram, the circle has a diameter of 12 in, and the diameters intersecting at *O* create vertical angles of 30°. What is the area of the shaded region?

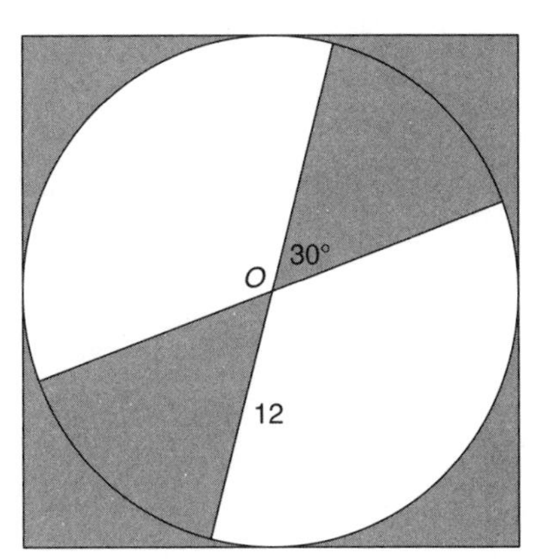

2. Each triangle in the following diagram is an isosceles right triangle. The large triangle has legs 30 cm long, and the small triangles have legs 10 cm long. Find the area of the shaded region.

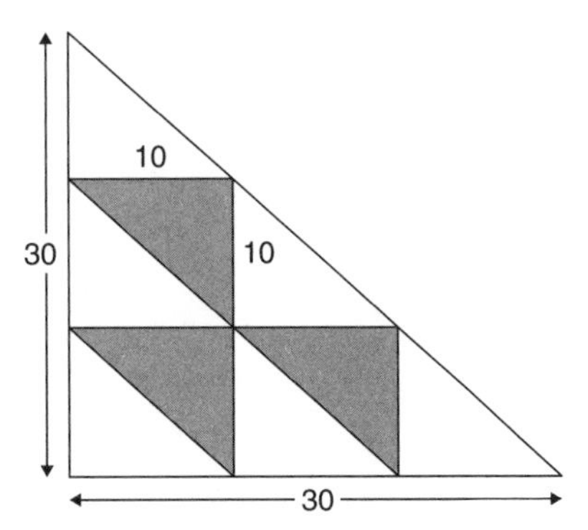

3. In the following diagram, the larger base of the white trapezoid is one-half the base of the parallelogram, and the shorter base of the trapezoid is one-third the base of the parallelogram. What fraction of the parallelogram is shaded?

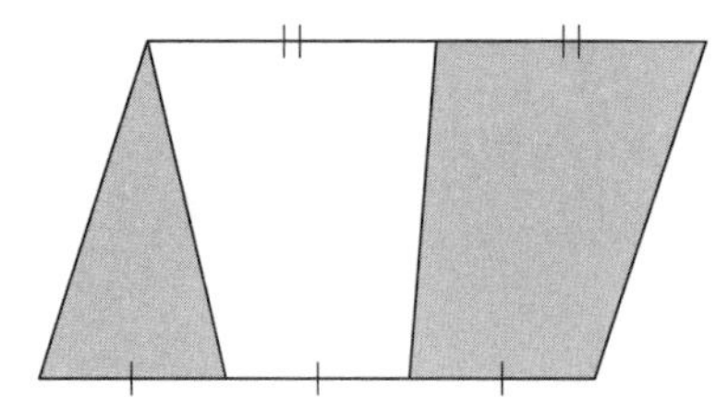

4. In the following diagram, the smallest of the four concentric circles has a radius of 3 cm, and each radius is 2 cm larger than the previous one. Find the area of the shaded region.

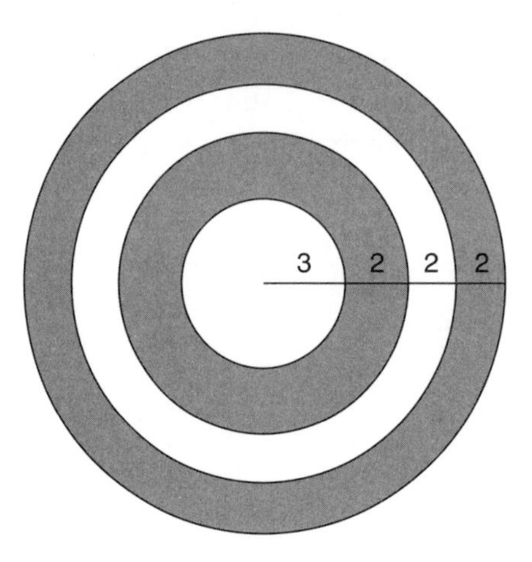

5. A regular hexagon is inscribed in a circle, as shown in the following diagram. If each side of the regular hexagon measures 8 cm, find the area of the shaded region.

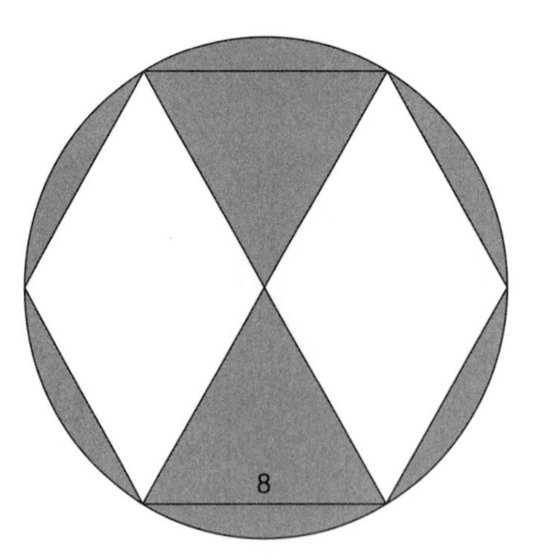

6. In the following diagram, the outer square has a side of 24 cm, and the inner square has a side of 12 cm. Find the area of the shaded region.

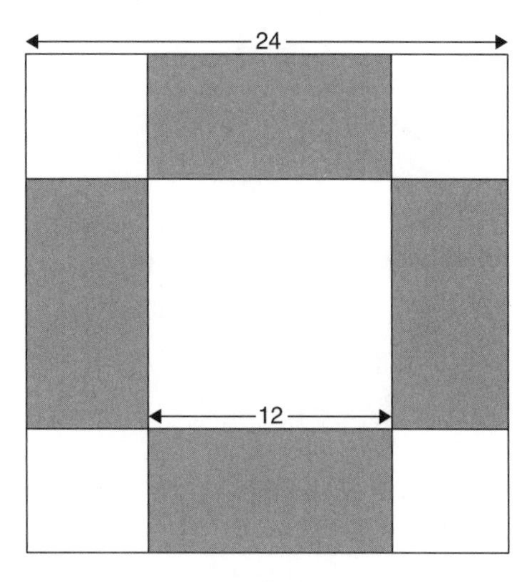

7. At its widest point, the following diagram measures 24 cm, and the diameters of the unshaded semicircles are 12 cm each. Find the area of the shaded region.

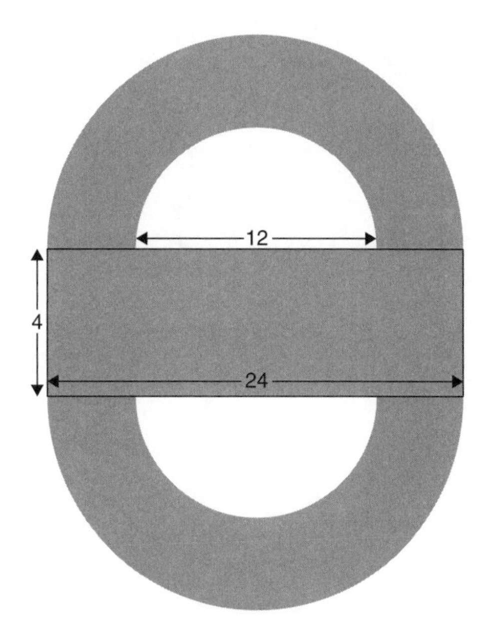

8. Find the area of the shaded region in the following diagram if the radius of the circle is 8 in, and the sectors are of equal size.

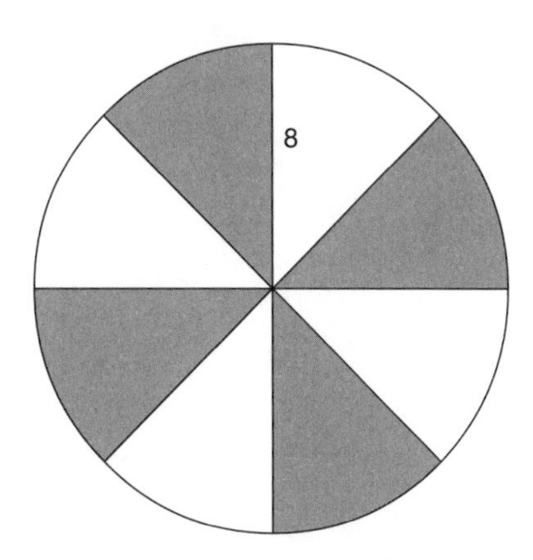

9. The regular pentagon that follows has a side of 15 cm. The inner regular pentagon has a side of 5 cm. Find the area of the shaded region.

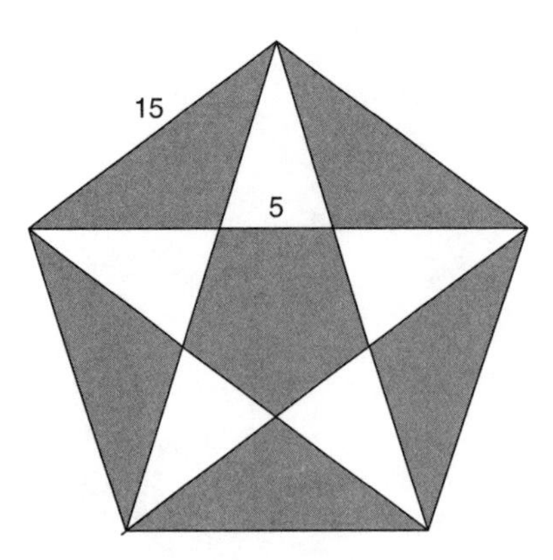

10. If the diameter of the following circle is 14 cm and the center square has a side of 2 cm, find the area of the shaded region.

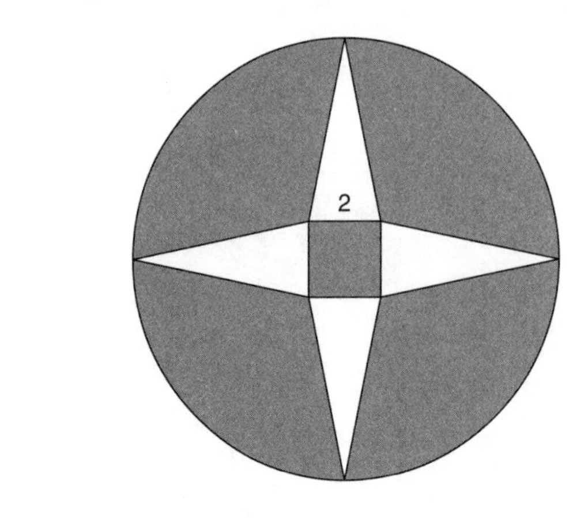

Surface area

·14·

A *polyhedron* is a three-dimensional figure in which all faces are polygons. The line segments where the polygons meet are the edges of the polyhedron. The surface area of a polyhedron is the total of the areas of all the faces.

Prisms

A *prism* is a polyhedron with two congruent parallel bases and sides that are parallelograms or rectangles. Figure 14.1 shows a hexagonal prism. The surface area of a prism is the sum of the areas of the two congruent parallel bases and the areas of the rectangles or parallelograms that are the lateral faces. Expressed as a formula, the surface area $S = 2B + Ph$, where S is the surface area, B is the area of the base polygon, P is the perimeter of the base polygon, and h is the height of the prism. The $2B$ term includes the areas of the two parallel bases. The type of polygon that serves as the base will determine how you calculate B. The total area of the parallelograms around the sides, the lateral area, is found by multiplying base by height for each face and then adding them. When you do that calculation, you get an expression that's equal to the perimeter of the base times the height, the Ph term.

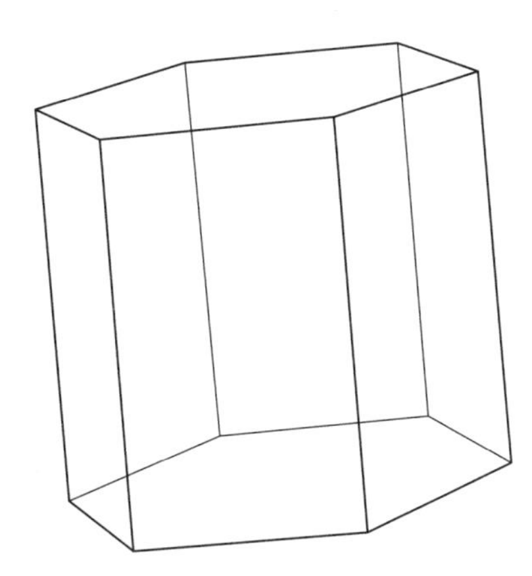

Figure 14.1 A hexagonal prism.

If the prism is a rectangular prism, the area of the base rectangle can be expressed as length times width, and two of the lateral faces have areas of length times height, while the other two are width times height. The surface area formula for a rectangular prism therefore becomes $S = 2lw + 2lh + 2wh$.

EXERCISE

14·1

Find the surface area of the prism described.

1. A regular octagonal prism 3 in high, with base area of 48 in^2 and perimeter $8\sqrt{10}$
2. A regular hexagonal prism 6 cm high, with base edge of 8 cm
3. A triangular prism 8 cm high, with base that is an equilateral triangle 4 cm on each side
4. A regular pentagonal prism 4.5 in high, with a side of 11 in and an apothem of 8 in
5. A right rectangular prism 1.5 m high, with rectangular base 2 m wide and 3 m long

Find the missing dimension of the prism described.

6. The total surface area of a triangular prism is 108 cm^2. If the base is a right triangle with legs of 3 and 4 cm, find the height of the prism.
7. The total surface area of a square prism is 406 cm^2. If the height of the prism is 11 cm, find the edge of the square base.
8. The total surface area of a regular hexagonal prism is $864\sqrt{3}$ cm^2. If the radius of the hexagonal base is 16 cm, find the height of the prism.
9. The total surface area of a regular pentagonal prism is 740 cm^2. If the apothem of the base is 5.5 cm and the height of the prism is 13 cm, find the perimeter of the pentagonal base to the nearest centimeter.
10. The total surface area of a regular octagonal prism is 6,232 cm^2. If the area of the base is 1,748 cm^2 and the base edge is 19 cm, find the height of the prism.

Cylinders

A *cylinder*, like the one in Figure 14.2, is a three-dimensional figure with identical circles as parallel bases. The lateral surface of the cylinder unrolls to a rectangle whose base is the circumference of the circle and whose height is the height of the cylinder. The surface area of a cylinder is $S = 2\pi r^2 + 2\pi rh$, where S is the surface area, r is the radius of the circular base, and h is the height of the cylinder. The $2\pi r^2$ gives the area of the two circles, and the $2\pi rh$ is circumference ($2\pi r$) times height for the lateral area.

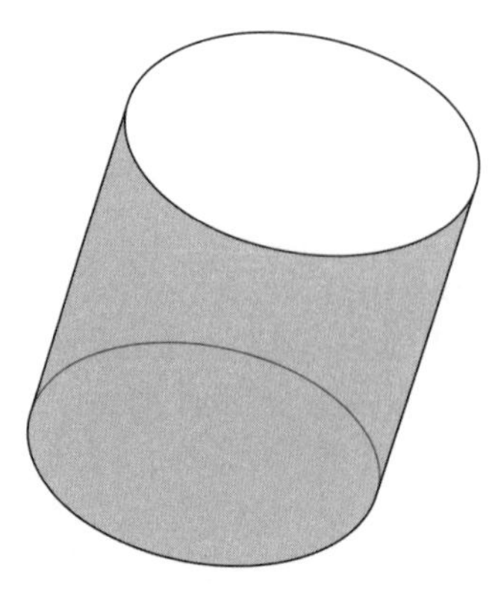

Figure 14.2

EXERCISE

14·2

Find the surface area of the cylinder with the given dimensions.

1. Radius = 12 cm, height = 30 cm
2. Diameter = 18 in, height = 7 in
3. Radius = 4 in, height = 9 in
4. Circumference = 38π cm, height = 24 cm
5. Radius = 5 in, height = 7 in

Find the indicated dimension of the cylinder described.

6. The total surface area of a cylinder is 1116π in^2. If the radius is 18 in, find the height.
7. The total surface area of a cylinder is 156π m^2. If the height is 7 m, find the radius.
8. The total surface area of a cylinder is $2{,}296\pi$ cm^2. If the diameter is 56 cm, find the height.
9. The total surface area of a cylinder is 748π ft^2. If the circumference of the base is 22π ft, find the height.
10. The total surface area of a cylinder is 100π m^2. If the combined area of the two bases is equal to the lateral area, find the diameter.

Pyramids

A *pyramid* has a single base, which is a polygon, and lateral faces that are triangles that meet at a single point. Figure 14.3 shows a hexagonal pyramid. Each of the triangular faces has a base that is equal to a side of the base polygon, but the height of the triangle is not the height of the pyramid. The altitude of the triangle is called the *slant height* of the pyramid, and it can be found by using the Pythagorean theorem if the height of the pyramid and the apothem of the polygonal base are known.

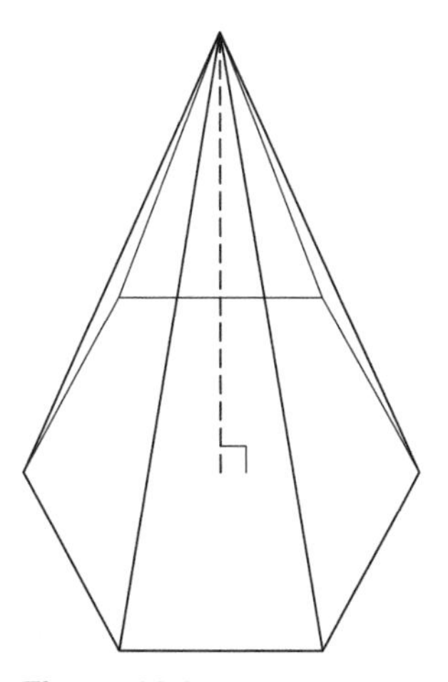

Figure 14.3

The surface area of a pyramid is the area of the polygon at the base plus the areas of the triangles that are the lateral faces. The area of each of these triangles is one-half the product of the base edge and the slant height. Taken together, they give a lateral area of $\frac{1}{2}Pl$, where P is the perimeter of the base and l is the slant height. The total surface area is $S = B + \frac{1}{2}Pl$, where S is the surface area, B is the area of the base, P is the perimeter of the base, and l is the slant height.

Find the slant height of the pyramid described.

1. A square pyramid 8 in high, with base edge of 12 in
2. A regular hexagonal pyramid 12 cm high, with base edge of 8 cm
3. A regular octagonal pyramid 35 ft high, if the apothem of the octagon measures 12 ft

Find the surface area of the pyramid.

4. A square pyramid with base edge of 2 m and a slant height of 3 m
5. A regular hexagonal pyramid with base edge of 20 in and a slant height of $5\sqrt{3}$ in
6. A triangular pyramid, with a base that is an equilateral triangle 10 cm on each side, if the slant height is $1\frac{2}{3}$ cm
7. A pentagonal pyramid if the perimeter of the base is 36 in, the apothem is 5 in, and the pyramid is 1 ft high. Find the missing dimension.
8. The total surface area of a triangular pyramid is $405\sqrt{3}$ in^2. If the base is an equilateral triangle 18 in on each side, find the slant height of the pyramid.
9. The total surface area of a square pyramid is 2,028 cm^2. If the perimeter of the base is 104 cm, find the slant height of the pyramid.
10. The total surface area of a pentagonal pyramid is 2,625 cm^2. If the perimeter of the base is 140 cm and the slant height is 33 cm, find the area of the base.

Cones

A *cone* is a three-dimensional figure with a circular base and a lateral surface that tapers to a point. The slant height l, the radius of the base r, and the height of the cone h form a right triangle with the slant height as the hypotenuse and the radius and height as legs. The three measurements are related by the Pythagorean theorem $h^2 + r^2 = l^2$, so if two of the measurements are known, the third can be calculated.

The surface area of the cone is $S = \pi r^2 + \pi rl$, where r is the radius of the base and l is the slant height. The πr^2 term is the area of the circular base, and the πrl term is the lateral area. The πrl comes from "unrolling" the lateral area and laying it flat. The result is a portion of a circle, and the radius of that circle is l, the slant height. The length of its arc is some fraction of the circumference $2\pi l$, and its area is the same fraction of πl^2, but what fraction? Well, the arc length has to be just long enough to go around the base circle, so $\frac{x}{360}\cdot 2\pi l = 2\pi r$. Do a little algebra and you'll find out that $x/360 = r/l$. So our flattened-out lateral area is

$$\frac{x}{360}\cdot \pi l^2 = \frac{r}{l}\cdot \pi l^2 = \pi rl$$

Find the slant height of the cone to the nearest tenth.

1. Radius of 14 cm and height of 10 cm
2. Height of 20 in and diameter of 16 in
3. Circumference of the base is 30π cm and height is 36 cm.

Find the surface area of the cone.

4. Radius of 3 ft and height of 4 ft
5. Diameter of 28 cm and slant height of 45 cm
6. Circumference of the base is 88π in and the slant height is 50 in.
7. Radius of 75 cm and height of 180 cm

Find the missing dimension.

8. The total surface area of a cone is 330π cm^2. If the radius of the base is 11 cm, find the slant height of the cone.
9. The total surface area of a cone is 42π in^2. If the slant height is 19 in, find the radius of the base.
10. The total surface area of a cone is 96π cm^2. If the diameter of the base is 12 cm, find the height of the cone.

Spheres

A *sphere* is the set of all points in space at a fixed distance from a given point. Most people would think of a sphere as a ball, even though most of the balls we use in sports are seamed or pieced together somehow rather than being the one smooth surface that is a sphere. The image is close enough for most problems, though, as long as we agree not to consider footballs. The surface area of a sphere is $S = 4\pi r^2$, where S is the surface area and r is the radius of the sphere.

EXERCISE
14·5

Find the surface area of the sphere with the given dimensions.

1. Radius = 3 ft
2. Diameter = 22 in
3. Radius = 6 m
4. Diameter = 36 cm
5. Radius = 15 in

Find the radius of the sphere with the given surface area.

6. Surface area = 144π ft^2
7. Surface area = $2{,}500\pi$ cm^2
8. Surface area = $1{,}444\pi$ cm^2
9. Surface area = 256π m^2
10. Surface area = 36π in^2

Surface areas of similar solids

If a pair of three-dimensional figures have corresponding measurements that are in proportion, they are similar solids. The ratio of corresponding sides is the scale factor. When two solids are similar with a scale factor of $a:b$, the ratio of their surface areas is $a^2 : b^2$.

Recall that the ratio of perimeters of similar polygons was equal to the scale factor, because perimeters add up sides, but the ratio of the areas of similar polygons was the scale factor squared, because each of the dimensions is multiplied by the scale factor and then multiplied together. The surface area is the sum of the areas of the faces. You're adding areas, so you're still in the world of scale factor squared.

1. Two triangular prisms are similar with a scale factor of 3:5. If the surface area of the smaller prism is 387 cm^2, find the surface area of the larger.
2. Two cones are similar with a scale factor of 12:25. If the surface area of the larger cone is 6,875 cm^2, find the surface area of the smaller.
3. Two square pyramids are similar. The corresponding sides measure 20 and 30 cm, and the surface area of the smaller pyramid is 510 cm^2. Find the surface area of the larger.
4. Two similar cylinders have surface areas of 304 and 912 cm^2. If the height of the smaller cylinder is 4 cm, find the height of the larger.
5. Two similar hexagonal prisms are similar with surface areas of 189 and 1,512 cm^2. If the base edge of the larger prism is 8 cm, find the corresponding edge of the smaller.
6. Two triangular pyramids are similar with a scale factor of 2:9. If the surface area of the smaller prism is 124 cm^2, find the surface area of the larger.

7. A cube is enlarged proportionally so that each edge increases by 3 in, and the surface area increases from 24 to 150 in^2. Find the edge of the original cube.
8. The surface areas of two similar cylinders are in ratio 81 : 16. If the radius of the smaller cylinder is 12 cm, find the circumference of the larger cylinder.
9. When a cube is enlarged proportionally so that each edge increases by 2 in, the surface area is multiplied by a factor of 4. Find the edge of the original cube.
10. A cone is reduced in size proportionally so that the new height is 5 cm less than the original, and the new surface area is one-fourth of the original. Find the height of the original cone.

Volume

The *volume* of a solid is a measurement of the space enclosed by, or occupied by, the solid. You might think of it as the amount a hollow three-dimensional figure could hold or the amount of water it would displace if it were submerged.

Prisms

The rule for finding the volume of a prism is the same no matter what polygon forms the base. The volume is the area of the base times the height of the prism. Of course, how you find the area of the base will depend on the polygon that forms the base. Volume of a prism $= Bh$, where B is the area of the base and h is the height of the prism.

Find the volume of the prism described.

1. A square prism 4 in high with square base 2 in on a side
2. A regular octagonal prism 3 in high, with base area of 48 in^2
3. A regular hexagonal prism 6 cm high, with base edge of 8 cm
4. A triangular prism 8 cm high, with base that is an equilateral triangle 4 cm on each side
5. A regular pentagonal prism 4.5 in high, with a side of 11 in and an apothem of 8 in
6. A rectangular prism 1.5 m high, with rectangular base 2 m wide and 3 m long

Find the missing measurement.

7. The volume of a right triangular prism is 48 cm^3. If the base is a right triangle with legs of 3 and 4 cm, find the height of the prism.
8. The volume of a right square prism is 539 cm^3. If the height of the prism is 11 cm, find the edge of the square base.
9. The volume of a regular hexagonal prism is 9,937.6 cm^3. If the radius of the hexagonal base is 15 cm, find the height of the prism to the nearest centimeter.
10. The volume of a regular pentagonal prism is 1,430 cm^3. If the apothem of the base is 5.5 cm and the height of the prism is 13 cm, find the edge of the pentagonal base.

Cylinders

Like prisms, cylinders have a volume equal to the area of the base times the height. In a cylinder, the base is a circle so the formula for the volume of a cylinder becomes $\pi r^2 h$, where r is the radius of the circle and h is the height of the prism.

Find the volume of the cylinder.

1. Radius of 3 cm, height of 12 cm
2. Radius of 12 cm, height of 30 cm
3. Diameter of 18 in, height of 7 in
4. Radius of 4 in, height of 9 in
5. Circumference of 38π cm, height of 24 cm

Find the missing measurement.

6. The volume of a cylinder is $4{,}212\pi$ in^3. If the radius is 18 in, find the height.
7. The volume of a cylinder is 252π m^3. If the height is 7 m, find the radius.
8. The volume of a cylinder is $10{,}192\pi$ cm^3. If the diameter is 56 cm, find the height.
9. The volume of a cylinder is $2{,}783\pi$ ft^3. If the circumference of the base is 22π ft, find the height.
10. The volume of a cylinder is $1{,}000\pi$ m^3. If the radius of the base equals the height, find the diameter.

Pyramids

Because pyramids taper to a point, the volume of a pyramid is much smaller than the volume of a prism with a matching base and height. The volume of a pyramid is actually only one-third the volume of a prism with the same base and height. $V = \frac{1}{3}Bh$, where B is the area of the base and h is the height of the pyramid.

Find the volume of the pyramid with the given dimensions.

1. A triangular pyramid 18 cm high with a base that is an equilateral triangle 6 cm on a side
2. A square pyramid 3 m high, with base edge of 2 m
3. A regular hexagonal pyramid 11 in high, with base edge of 20 in
4. A triangular pyramid, with a base that is an equilateral triangle 10 cm on each side, if the pyramid is 12 cm high
5. A pentagonal pyramid if the perimeter of the base is 35 in, the apothem is $5\frac{1}{4}$ in, and the pyramid is 1 ft high

6. A triangular pyramid 2 ft high with a base that is an equilateral triangle 4 ft on a side

Find the missing measurement.

7. The volume of a triangular pyramid is $640\sqrt{3}$ in³. If the edge of the equilateral triangle base is 16 in, find the height of the pyramid.
8. The volume of a square pyramid is 2,916 cm³. If the perimeter of the base is 108 cm, find the height of the pyramid.
9. The volume of a pentagonal pyramid is 2,827 cm³. If the height is 33 cm, find the area of the base.
10. A pyramid with a regular hexagon as its base is 8 in high. If its volume is $432\sqrt{3}$ in³, find the edge of the base.

Cones

Because the cone, like the pyramid, tapers to a point, its volume is one-third of the volume of a cylinder with the same radius and height. $V = \frac{1}{3}\pi r^2 h$, where V is the volume of the cone, r is the radius of the base, and h is the height of the cone.

Find the volume of the cone.

1. Radius of 12 cm and height of 18 cm
2. Radius of 3 ft and height of 4 ft
3. Diameter of 28 cm and height of 45 cm
4. Circumference of the base of 84π in and height of 50 in
5. Radius of 75 cm and height of 180 cm
6. Radius of 15 cm and height of 10 cm

Find the missing dimension.

7. The volume of a cone is 605π cm³. If the radius of the base is 11 cm, find the height of the cone.
8. The volume of a cone is 18π in³. If the height is 2 in, find the radius of the base.
9. The volume of a cone is 48π cm³. If the diameter of the base is 8 cm, find the height of the cone.
10. A wooden cone is packaged in a cardboard box that is a square prism, and the empty space is filled with packing material. If the diameter of the cone and the side of the square both measure 4 in and the height of the cone and the height of the prism both measure 12 in, find the volume of packing material needed, to the nearest cubic inch.

Spheres

The volume of a sphere is $\frac{4}{3}$ times π times the cube of the radius. The formula for the volume of a sphere is difficult to explain without using calculus, but if you remember that the area of a circle is πr^2, it doesn't seem too surprising to find πr^3 in the formula for the volume of a sphere. The $\frac{4}{3}$ is the part that seems surprising, but until you take calculus, you'll have to just trust that it's correct. $V = \frac{4}{3}\pi r^3$, where V is the volume of the sphere and r is its radius.

Find the volume of the sphere with the given radius or diameter.

1. Radius = 9 in
2. Radius = 3 ft
3. Diameter = 22 in
4. Radius = 6 m
5. Diameter = 36 cm
6. Radius = 15 in

Find the radius of the sphere with the given volume.

7. Volume = 36π in^3
8. Volume = 288π in^3
9. A spherical ball with a radius of 2 ft is coated with an even layer of plastic until the radius of the coated sphere is 27 in. Find the volume of the plastic layer.
10. The shape of a silo is a cylinder topped by a hemisphere. Assume that there is no barrier between the hemispherical section and the cylindrical section. If the structure is 20 ft in diameter and 110 ft high, what is the volume of the silo?

Volumes of similar solids

If two solids are similar—that is, their corresponding measurements are in proportion—with a scale factor of $a:b$, then the ratio of their surface areas is $a^2:b^2$ and the ratio of their volumes is $a^3:b^3$. We jump to the third power of the scale factor when we get to volume because calculating volume involves multiplying three dimensions, each of which is multiplied by the scale factor.

1. Two triangular prisms are similar with a scale factor of 3:5. If the volume of the smaller prism is 216 cm^3, find the volume of the larger.
2. Two cones are similar with a scale factor of 1:2. If the volume of the larger cone is 280 cm^3, find the volume of the smaller.
3. Two square pyramids are similar. The corresponding sides measure 20 and 30 cm, and the volume of the smaller pyramid is 10,200 cm^3. Find the volume of the larger.

4. Two similar cylinders have volumes of 1,216 and 6,517 cm^3. If the height of the smaller cylinder is 4 cm, find the height of the larger.
5. Two hexagonal prisms are similar with volumes of 567 and 10,752 cm^3. If the base edge of the larger prism is 8 cm, find the corresponding edge of the smaller.
6. Two triangular pyramids are similar with a scale factor of 2:9. If the volume of the smaller prism is 248 cm^3, find the volume of the larger.
7. Two cubes are similar with a scale factor of 3:5. The larger cube has a volume of 1,500 cm^3. What is the volume of the smaller cube?
8. The volumes of two similar cylinders are in ratio 2,187:192. If the radius of the smaller cylinder is 12 cm, find the circumference of the larger cylinder.
9. When a cube is enlarged proportionally so that each edge increases by 2 in, the volume is multiplied by a factor of 8. Find the edge of the original cube.
10. A cone is reduced in size proportionally so that the new height is 5 cm less than the original, and the new volume is one-eighth of the original. Find the height of the original cone.

Answer key

1 Logic and reasoning

1·1

1. 19
2. $\frac{9}{10}$
3. 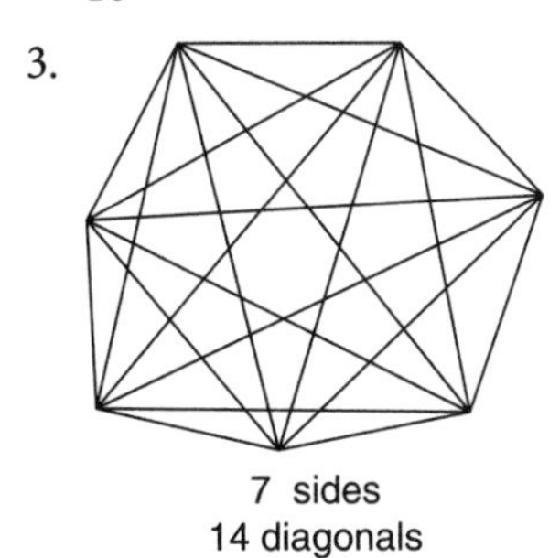

4. 25
5. □
□□
□□□
□□□□
□□□□□
6. The squares of odd numbers are odd. A counterexample would have to be an odd number whose square is even, but that doesn't exist.
7. The diagonals of any rectangle are equal to one another. A counterexample would have to be a rectangle with diagonals of different lengths, but you'll never find one.
8. If you wear a green shirt, there will be a pop quiz in English. Counterexample: You wear a green shirt and there is no pop quiz.
9. The sum of the lengths of the two shorter sides of a triangle is greater than the length of the longest side. A counterexample would have to be a triangle in which two sides add to less than the third side, but that's impossible since the straight-line distance between two points is always the shortest.
10. Each power of 2 is one more than the sum of the powers before it. A counterexample would be a power of 2 that was more or less than 1 more than the sum of the preceding powers (but you won't find one).

1·2

1. $p \wedge r$: A touchdown scores 6 points, and you can kick for an extra point after a touchdown. (True)
2. $q \vee t$: A field goal scores 3 points or a fumble scores 10 points. (True)
3. $\sim r \vee p$: You cannot kick for an extra point after a touchdown or a touchdown scores 6 points. (True)
4. $q \wedge \sim t$: A field goal scores 3 points and a fumble does not score 10 points. (True)

5. $\sim p \vee \sim q$: A touchdown does not score 6 points or a field goal does not score 3 points. (False)
6. $t \wedge \sim p$: A fumble scores 10 points and a touchdown does not score 6 points. (False)
7. $\sim p \wedge \sim t$: A touchdown does not score 6 points and a fumble does not score 10 points. (False)
8. $q \vee \sim r$: A field goal scores 3 points or you cannot kick for an extra point after a touchdown. (True)
9. $\sim r \wedge \sim t$: You cannot kick for an extra point after a touchdown, and a fumble does not score 10 points. (False)
10. $p \vee q \vee \sim r$: A touchdown scores 6 points or a field goal scores 3 points or you cannot kick for an extra point after a touchdown. (True)
11.

p	q	$\sim p$	$\sim p \vee q$
T	T	F	T
T	F	F	F
F	T	T	T
F	F	T	T

12.

p	q	$\sim p$	$\sim q$	$\sim p \wedge \sim q$
T	T	F	F	F
T	F	F	T	F
F	T	T	F	F
F	F	T	T	T

13.

p	q	$\sim q$	$p \wedge \sim q$
T	T	F	F
T	F	T	T
F	T	F	F
F	F	T	F

14.

p	q	$p \wedge q$	$\sim(p \wedge q)$
T	T	T	F
T	F	F	T
F	T	F	T
F	F	F	T

15.

p	q	$\sim p$	$\sim q$	$\sim p \vee \sim q$
T	T	F	F	F
T	F	F	T	T
F	T	T	F	T
F	F	T	T	T

1·3

1. *C*: If wood turns to ash, then it is burned. *I*: If wood is not burned, then it doesn't turn to ash. *CP*: If wood does not turn to ash, then it is not burned.
2. *C*: If a team kicks an extra point, then they scored a touchdown. *I*: If a team doesn't score a touchdown, then they don't kick an extra point. *CP*: If a team doesn't kick an extra point, then they don't score a touchdown.
3. *C*: If you don't score well on your exam, then you didn't study. *I*: If you study for your exam, then you'll score well. *CP*: If you score well on your exam, then you studied.
4. *C*: If you're hungry in the afternoon, then you didn't eat lunch. *I*: If you eat lunch, then you won't be hungry in the afternoon. *CP*: If you're not hungry in the afternoon, then you ate lunch.
5. *C*: If you're tired, then you ran a marathon. *I*: If you don't run a marathon, then you won't be tired. *CP*: If you're not tired, then you didn't run a marathon.
6. A number is odd if and only if it is 1 less than an even number. (True)
7. A number is divisible by 3 if and only if it is divisible by 6. (False)
8. A number is a multiple of 3 if and only if its digits add to a multiple of 3. (True)
9. A triangle is equilateral if and only if it is equiangular. (True)
10. The first day of the month is a Tuesday if and only if the 30th day of the month is a Thursday. (False)
11.

p	q	$\sim p$	$\sim p \rightarrow q$
T	T	F	T
T	F	F	T
F	T	T	T
F	F	T	F

12.

p	q	$\sim p$	$\sim q$	$\sim p \wedge \sim q$	$(\sim p \wedge \sim q) \rightarrow q$
T	T	F	F	F	T
T	F	F	T	F	T
F	T	T	F	F	T
F	F	T	T	T	T

13.

p	q	$\sim q$	$p \wedge \sim q$	$(p \wedge \sim q) \to p$
T	T	F	F	T
T	F	T	T	T
F	T	F	F	T
F	F	T	F	T

14.

p	q	$\sim p$	$\sim q$	$p \wedge q$	$\sim(p \wedge q)$	$\sim p \vee \sim q$	$[\sim(p \wedge q) \leftrightarrow [\sim p \vee \sim q]$
T	T	F	F	T	F	F	T
T	F	F	T	F	T	T	T
F	T	T	F	F	T	T	T
F	F	T	T	F	T	T	T

15.

p	q	$\sim p$	$\sim q$	$p \vee q$	$\sim(p \vee q)$	$\sim p \wedge \sim q$	$[\sim(p \vee q)] \leftrightarrow [\sim p \wedge \sim q]$
T	T	F	F	T	F	F	T
T	F	F	T	T	F	F	T
F	T	T	F	T	F	F	T
F	F	T	T	F	T	T	T

1·4

1. Everyone is leaving the building.
2. If you study geometry, then your essays will be better structured.
3. No conclusion.
4. If you practice an instrument every day, then you may have a career in music.
5. No conclusion.
6. Invalid. Reasoning from the inverse.
7. Valid
8. Invalid. While the conclusion is a reasonable statement, it does not follow logically from the previous statements. If I is "you practice an instrument every day," P is "you will improve your playing" and C is "you want a career in music," the structure of the argument is $I \to P$ and $C \to P$, which don't connect as a syllogism to allow you to conclude $C \to I$.
9. Invalid. If H is "you do your math homework," T is "you will pass the math test" and V is "you'll be able to play video games," then the structure of the argument is $H \to T$ and $H \to V$. This doesn't allow you to conclude that $V \to T$.
10. Valid. (Reasoning from the contrapositive.)

1·5

1. Addition property of equality
2. Division property of equality
3. Multiplication property of equality
4. Subtraction property of equality, commutative property for addition
5. Associative property for addition
6. Subtraction property of equality, commutative property for addition
7. Substitution
8. Distributive property
9. $8(x-4)-16=10(x-7)$

$(8x-32)-16=10x-70$	Distributive property
$8x+(-32-16)=10x-70$	Associative property for addition
$8x-48=10x-70$	Substitution
$8x-48-8x=10x-70-8x$	Subtraction property of equality
$8x-8x-48=10x-8x-70$	Commutative property for addition
$-48=2x-70$	Substitution
$-48+70=2x-70+70$	Addition property of equality
$22=2x$	Substitution
$\frac{22}{2}=\frac{2x}{2}$	Division property of equality
$11=x$	Substitution

10. $8(2x-5)-2(x-2)=5(x+7)-4(x+8)$

$(16x-40)+(-2x+4)=(5x+35)+(-4x-32)$	Distributive property
$(16x-2x)+(-40+4)=(5x-4x)+(35-32)$	Commutative and associative properties
$14x-36=x+3$	Substitution
$14x-36-x=x+3-x$	Subtraction property of equality
$14x-x-36=x-x+3$	Commutative property
$13x-36=3$	Substitution
$13x-36+36=3+36$	Addition property of equality
$13x=39$	Substitution
$\frac{13x}{13}=\frac{39}{13}$	Division property of equality
$x=3$	Substitution

1·6

1. If we have a line and a point not on that line, and we know a line contains at least two points, then we have at least three points, and they are not all on the same line. Three points not all on the same line define a unique plane that contains them.
2. It's hard to prove this directly, so prove the contrapositive instead. If the contrapositive is true, the statement is also true. The contrapositive says, "If they intersect in more than one point, then it is not true that two lines intersect." If two lines do intersect in more than one point, then there are at least two points that lie on both lines. We know that two points determine one and only one line, however, so there are not two intersecting lines, but only one line.
3. If we have two lines and they intersect, then they intersect in only one point. Each line contains at least two points, so there's another point (besides the intersection point) on each line. One additional point on each line plus the intersection point gives us at least three points.
4. If two lines intersect, then the intersection point and one point chosen from each line are three points, not all on the same line. These points determine a unique plane that contains them. Since this plane contains two points from each line, it must also contain both lines. It is the only plane containing the three points, so it is also the only plane containing both lines.
5. Counterexample: Skew lines do not intersect, but are in different planes. In the following diagram, $\overleftrightarrow{QR}$ will never intersect $\overleftrightarrow{XY}$, but there is no plane that contains both of them.

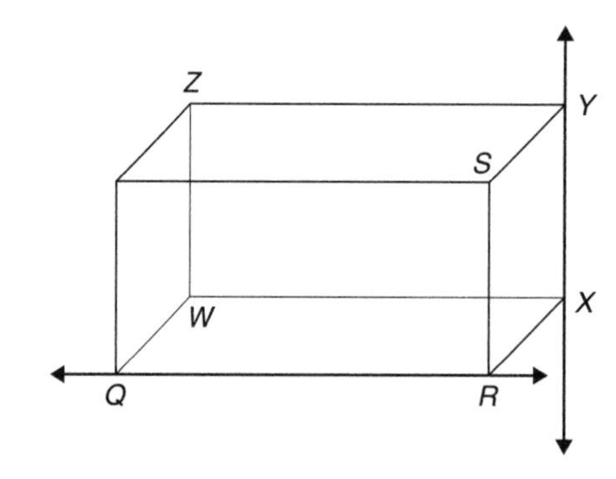

6. A line contains at least two points. If the two lines do not intersect, they do not share any points, so there are at least four points. Since the plane contains both lines, it contains the four or more points on those lines.
7. Counterexample: The distance from A to C plus the distance from C to B in the following diagram is much bigger than the distance from A to B. The statement is true only if C is between A and B.

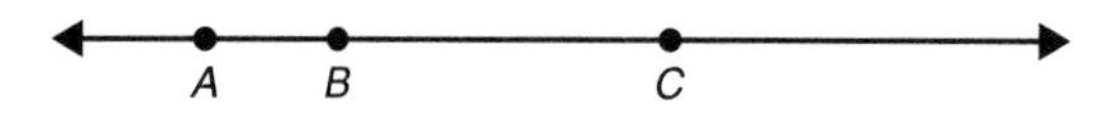

8. Counterexample: As the preceding diagram shows, it is possible to arrange points A, B, and C on a line so that the distance from A to C plus the distance from C to B is greater than the distance from A to B.
9. If a plane contains a line, there are at least two points on the line, and therefore on the plane. We know that a plane contains at least three points, not all on the same line, so there is at least one other point on the plane that is not on the line. Those points, together with either one of the points on the line, determine another line.
10. If two planes have a point in common, then they intersect; and if two planes intersect, their intersection is a line. A line contains at least two points, so the intersection of the two planes is at least two points.

2 Lines and angles

2·1

1. Points S and T
2. Line m or line t
3. Plane P or plane Q
4. Line l or line m
5. Plane P
6. Point T
7. Line l
8. Line m or line l
9. Line m
10. Plane Q

2·2

1. $\overrightarrow{SP}$, $\overrightarrow{ST}$, or $\overrightarrow{SN}$
2. $\overrightarrow{XL}$ and $\overrightarrow{XY}$
3. $\overrightarrow{SP}$
4. $\overrightarrow{TM}$ and $\overrightarrow{TN}$ or $\overrightarrow{TM}$ and $\overrightarrow{TP}$
5. $\overrightarrow{SN}$
6. False
7. True
8. $\overline{XY}$ or XR or XL
9. $\overline{QR}$ and $\overline{QW}$. Any two of QR, QW, QT, or QM.
10. True

2·3

1. 12
2. 6
3. 10
4. 2
5. 15
6. 9
7. 4
8. 8 or −8
9. 8 or −2
10. −2 or −10

2·4

1. $\angle PRW$ or $\angle QRW$ or $\angle PRV$ or $\angle QRV$
2. $\angle WVS$ or $\angle WVZ$
3. $\angle PRW$ ($\angle 1$), $\angle TRW$ ($\angle 2$) and $\angle PRT$
4. $\angle PQX$
5. $\angle 3$
6. Y
7. $\overrightarrow{RW}$ (or RV) and $\overrightarrow{RT}$ (or RS)
8. True
9. False
10. False

2·5

1. 120°
2. 75°
3. 30°
4. 50°
5. 60°
6. 95°
7. 115°
8. 100°
9. 20°
10. 85°
11. Acute
12. Straight
13. Acute
14. Right
15. Acute
16. Acute
17. Obtuse
18. Obtuse
19. Acute
20. Obtuse

2·6

1. P
2. −2
3. −5
4. −3
5. 10
6. False
7. False
8. $\overrightarrow{PR}$
9. 100
10. 50

2·7

1. 37°
2. 59°
3. 133°
4. 79°
5. $\angle ACG$ or $\angle FCL$ or $\angle ACL$ or $\angle FCG$
6. $\angle DAG$ or $\angle DAJ$
7. $\angle LBK$
8. $\angle GBA$
9. $\angle DAE$
10. $\angle CGA$

3 Parallel and perpendicular lines

3·1

1. $\angle RQW$
2. $\angle QTU$
3. $\angle PQW$
4. $\angle PQT$
5. $\angle VTU$
6. $\angle PQW$ or $\angle RQT$
7. $\angle STQ$
8. $\angle STQ$
9. $\angle RQW$
10. $\angle QTU$
11. 38°
12. 139°
13. 112°
14. 125°
15. 61°
16. 131°
17. 18°
18. 23°
19. 73°
20. 99°

3·2

1. Always
2. Never
3. Sometimes
4. Sometimes
5. Sometimes
6. Sometimes
7. Never
8. Never
9. Sometimes
10. Sometimes

3·3

1. $\overleftrightarrow{AB} \parallel \overleftrightarrow{CD}$. If two lines are cut by a transversal so that alternate exterior angles are congruent, then the lines are parallel.
2. No parallel lines
3. $\overleftrightarrow{AB} \parallel \overleftrightarrow{CD}$. If two lines are cut by a transversal so that corresponding angles are congruent, then the lines are parallel.
4. $\overleftrightarrow{AB} \parallel \overleftrightarrow{CD}$. If two lines are cut by a transversal so that interior angles on the same side of the transversal are supplementary, then the lines are parallel.
5. $\overleftrightarrow{EH} \parallel \overleftrightarrow{JM}$. If two lines are cut by a transversal so that alternate exterior angles are congruent, then the lines are parallel.
6. $\overleftrightarrow{AB} \parallel \overleftrightarrow{CD}$. If two lines are perpendicular to the same line, then they are parallel to each other.
7. No parallel lines
8. No parallel lines
9. $\overleftrightarrow{AB} \parallel \overleftrightarrow{CD}$. If two lines are cut by a transversal so that alternate interior angles are congruent, then the lines are parallel.
10. $\overleftrightarrow{EH} \parallel \overleftrightarrow{JM}$. If two lines are cut by a transversal so that alternate interior angles are congruent, then the lines are parallel.

4 Congruent triangles

4·1

1. $\overline{MI} \cong \overline{FR}, \overline{IX} \cong \overline{RU}, \overline{XE} \cong \overline{UI}, \overline{ED} \cong \overline{IT}, \overline{MD} \cong \overline{FT}, \angle M \cong \angle F, \angle I \cong \angle R, \angle X \cong \angle U, \angle E \cong \angle I, \angle D \cong \angle T$
2. $\overline{CA} \cong \overline{DO}, \overline{AT} \cong \overline{OG}, \overline{CT} \cong \overline{DG}, \angle C \cong \angle D, \angle A \cong \angle O, \angle T \cong \angle G$
3. $\overline{ST} \cong \overline{KT}, \overline{TA} \cong \overline{TE}, \overline{AR} \cong \overline{ER}, \overline{SR} \cong \overline{KR}, \angle S \cong \angle K, \angle T \cong \angle T, \angle A \cong \angle E, \angle R \cong \angle R$
4. $\overline{CA} \cong \overline{GA}, \overline{AN} \cong \overline{AR}, \overline{ND} \cong \overline{RD}, \overline{DL} \cong \overline{DE}, \overline{LE} \cong \overline{EN}, \overline{CE} \cong \overline{GN}, \angle C \cong \angle G, \angle A \cong \angle A, \angle N \cong \angle R, \angle D \cong \angle D, \angle L \cong \angle E, \angle E \cong \angle N$
5. $\overline{LI} \cong \overline{EA}, \overline{IP} \cong \overline{AR}, \overline{LP} \cong \overline{ER}, \angle L \cong \angle E, \angle I \cong \angle A, \angle P \cong \angle R$
6. $\Delta RST \cong \Delta GHI$
7. $LEAF \cong TEMS$
8. $GLOVE \cong SCARF$
9. $\Delta ARM \cong \Delta LEG$
10. $HOME \cong BASE$

4·2

1. $\Delta ABC \cong \Delta XYZ$, ASA or AAS
2. $\Delta CAT \cong \Delta DOG$, SSS
3. $\Delta BOX \cong \Delta CAR$, SAS
4. Cannot be determined
5. $\Delta ACT \cong \Delta WIN$, AAS
6. $\Delta BAG \cong \Delta ICE$, SAS
7. Cannot be determined
8. $\Delta ART \cong \Delta PEN$, AAS
9. Cannot be determined
10. $\Delta TOY \cong \Delta JAM$, SSS

4·3

1. $\Delta EAR \cong \Delta KIN$, SSS, $\angle E \cong \angle K$, $\angle A \cong \angle I$, $\angle R \cong \angle N$
2. $\Delta LIP \cong \Delta FEW$, SAS, $\angle L \cong \angle F$, $\angle P \cong \angle W$, $\overline{LP} \cong \overline{FW}$
3. $\Delta LEG \cong \Delta FOR$, ASA or SAS *or AAS*, $\angle L \cong \angle F$, $\overline{EL} \cong \overline{OF}$
4. $\Delta ARM \cong \Delta LOW$, AAS, $\angle A \cong \angle L$, $\overline{MR} \cong \overline{WO}$, $\overline{AR} \cong \overline{LO}$
5. $\Delta TOE \cong \Delta FAR$, SAS, $\angle T \cong \angle F$, $\angle O \cong \angle A$, $\overline{TO} \cong \overline{FA}$
6. $\Delta BOY \cong \Delta MEN$, SSS, $\angle B \cong \angle M$, $\angle O \cong \angle E$, $\angle Y \cong \angle N$
7. Cannot be determined
8. $\Delta CAN \cong \Delta SEW$, ASA, $\overline{CA} \cong \overline{SE}$, $\overline{AN} \cong \overline{EW}$, $\angle CAN \cong \angle SEW$
9. $\Delta DIM \cong \Delta TAP$, SAS, $\overline{IM} \cong \overline{AP}$, $\angle I \cong \angle A$, $\angle M \cong \angle P$
10. $\Delta GET \cong \Delta WHO$, SSS, $\angle G \cong \angle W$, $\angle E \cong \angle H$, $\angle T \cong \angle O$

4·4

1. $\Delta ABE \cong \Delta DCE$, SAS
2. $\Delta STX \cong \Delta VUX$, ASA *or AAS*
3. $\Delta DGE \cong \Delta FGC$, SAS
4. $\Delta ABF \cong \Delta CDE$ (SAS). $\overline{CE} \cong \overline{AF}$ (CPCTC). Prove $\Delta CAE \cong \Delta ACF$ (SAS).
5. $\Delta DYB \cong \Delta EYR$ (SAS). $\angle B \cong \angle R$, $\overline{BD} \cong \overline{RE}$ (CPCTC). Show $\overline{BE} \cong \overline{RD}$ by addition and prove $\Delta BDE \cong \Delta RED$ (SAS).
6. $\Delta HWA \cong \Delta FYC$ (SAS). $\overline{HA} \cong \overline{CF}$ (CPCTC). Prove $\Delta HAB \cong \Delta FCD$ (SSS).
7. $\Delta AED \cong \Delta BEC$ (SAS). $\overline{AD} \cong \overline{BC}$ (CPCTC). Prove $\Delta BAD \cong \Delta ABC$ (SAS *or SSS*).
8. $\Delta DGE \cong \Delta CGF$ (SAS – show $\overline{EG} \cong \overline{FG}$ and $\angle DGE \cong \angle CGF$). $\overline{DE} \cong \overline{CF}$, $\angle GDE \cong \angle GCF$ (CPCTC). Show $\angle DEA \cong \angle CFB$ and prove $\Delta ADE \cong \Delta BCF$(AAS). *Or show $\angle DEG = \angle GFC$ (CPCTC), so by addition and supplementary angles, $\angle AED = \angle CFB$. Prove $\Delta ADE \cong \Delta BCF$(AAS).*
9. $\Delta RXP \cong \Delta SXP$ (SAS). $\overline{RP} \cong \overline{SP}$ (CPCTC). Prove $\Delta PQR \cong \Delta PTS$ (SAS).
10. $\Delta WBX \cong \Delta YDZ$ (SSS). $\angle B \cong \angle D$, $\angle XWB \cong \angle ZYD$ (CPCTC). Show $\overline{CD} \parallel \overline{AB}$ ($\angle XWB$ and $\angle ZYD$ are a pair of corresponding angles and are congruent) and $\angle CAW \cong \angle ECY$ and prove $\Delta ABC \cong \Delta CDE$ (AAS).

4·5

1. State the given, and show that $\angle EBC \cong \angle ECB$. Show $\overline{AC} \cong \overline{BD}$ by adding BC to AB and CD. Prove $\Delta CAE \cong \Delta BDE$ by SAS. ($\overline{BE} \cong \overline{CE}$, $\angle EBC \cong \angle ECB$, $\overline{BD} \cong \overline{AC}$.)
2. State the given and show $\overline{SX} \cong \overline{XV}$ and $\overline{RX} \cong \overline{XW}$. Vertical angles are congruent, so $\angle SXR \cong \angle VXW$. Prove $\Delta SXR \cong \Delta VXW$ by SAS. Use CPCTC to show $\angle R \cong \angle W$. Then prove $\Delta TXR \cong \Delta UXW$ by ASA ($\angle R \cong \angle W$, $\overline{RX} \cong \overline{XW}$, $\angle TXR \cong \angle UXW$). $\angle T \cong \angle U$ by CPCTC.
3. State the given and show that $\angle DEF$ and $\angle CFE$ are right angles and therefore $\angle DEF \cong \angle CFE$. Prove $\Delta DEF \cong \Delta CFE$ by SAS ($\overline{ED} \cong \overline{FC}$, $\angle DEF \cong \angle CFE$, $\overline{EF} \cong \overline{EF}$). Show $\overline{DF} \cong \overline{EC}$ and $\angle EFD \cong \angle FEC$ by CPCTC. Then prove $\overline{AF} \cong \overline{BE}$ by adding EF to AE and BF. Prove $\Delta ADF \cong \Delta BCE$ by SAS.
4. State the given and prove $\Delta EDC \cong \Delta FBA$ (SAS). Show $\overline{EC} \cong \overline{FA}$ by CPCTC. Prove $\Delta ECA \cong \Delta FAC$ (SSS) and $\angle FAC \cong \angle ECA$ (CPCTC).
5. State the given, Prove $\Delta BYD \cong \Delta RYE$ (SAS). Then $\overline{YD} \cong \overline{YE}$ (CPCTC) and therefore $\angle YDE \cong \angle YED$. Prove $\Delta DEB \cong \Delta EDR$ (AAS).

5 Inequalities

5·1

1. Yes
2. Yes
3. No, $12 \text{ m} + 18 \text{ m} < 31 \text{ m}$
4. Yes
5. No, $5 \text{ mm} + 5 \text{ mm} = 10 \text{ mm}$
6. $3 \text{ cm} < c < 13 \text{ cm}$
7. $7 \text{ cm} < r < 15 \text{ cm}$
8. $9 \text{ cm} < z < 33 \text{ cm}$
9. $5 \text{ cm} < f < 43 \text{ cm}$
10. $11 \text{ cm} < t < 25 \text{ cm}$

5·2

1. $\overline{AC}, \overline{BC}, \overline{AB}$
2. $\overline{RT}, \overline{ST}, \overline{RS}$
3. $\overline{YZ}, \overline{XZ}, \overline{XY}$
4. $\overline{EF}, \overline{DE}, \overline{DF}$
5. $\overline{PQ}, \overline{PR}, \overline{QR}$
6. $\angle R, \angle Q, \angle P$
7. $\angle A, \angle B, \angle C$
8. $\angle Y, \angle X, \angle Z$
9. $\angle R, \angle T, \angle S$
10. $\angle E, \angle D, \angle F$

5·3

1. $CD < BC < BD < AB < AD$
2. $LK < IL < IK < IJ < JK$
3. $EH < EF < FH < FG < GH$
4. $NO < MN < MO < OP < MP$
5. $RS < RQ < QS < ST < QT$

5·4

1. $AC < DF$
2. $RS > MN$
3. $YZ < RT$
4. $QR > AB$
5. $LN = QP$
6. $CA < ED$
7. $XY = RT$
8. $GK = KH$
9. $ST > SR$
10. $MN < NO$

5·5

1. $m\angle B < m\angle E$
2. $m\angle S < m\angle Y$
3. $m\angle X > m\angle P$
4. $m\angle V < m\angle N$
5. $m\angle A > m\angle D$
6. $m\angle RVS = m\angle RVT$
7. $m\angle ADC > m\angle ADB$
8. $m\angle RSV < m\angle VST$
9. $m\angle BDC < m\angle BDA$
10. $m\angle KMJ < m\angle KML$

6 Quadrilaterals and other polygons

6·1

1. Quadrilateral, convex
2. Pentagon, convex
3. Dodecagon, concave
4. Triangle, convex
5. Octagon, concave
6. 9
7. 20
8. 35
9. 44
10. 171

6·2

1. 540°
2. 1080°
3. 720°
4. 1440°
5. 2880°
6. 135°
7. 150°
8. 120°
9. 162°
10. 108°
11. 60°
12. 73°
13. 24°
14. 121°
15. 9 sides

6·3

1. 96°
2. 88°
3. 90°
4. 99°
5. 36°

6·4

1. 96°
2. 145°
3. 73°
4. 115°
5. 107°
6. 20 cm
7. 22 in
8. 23 cm
9. 43 in
10. 281 mm

6·5

1. Parallelogram
2. Not a parallelogram
3. Parallelogram
4. Not a parallelogram
5. Parallelogram
6. $m\angle B = 128°$, $m\angle C = 52°$, and $m\angle D = 128°$
7. $m\angle B = 87°$, $m\angle C = 93°$, and $m\angle D = 87°$
8. $m\angle C = 107°$ and $m\angle D = 73°$
9. $m\angle B = 69°$ and $m\angle D = 69°$
10. $m\angle B = 71°$ and $m\angle D = 71°$
11. 12 cm
12. 9 in
13. 15 cm
14. 6 cm
15. 59 cm

6·6

1. Square
2. Parallelogram
3. Rectangle
4. Rhombus
5. Rhombus
6. $12\sqrt{2}$ cm
7. 12 in
8. 5 m
9. 24 cm
10. 10 m
11. 62°
12. 90°
13. 45°
14. 74°
15. 104°

6·7

1. Remember that to prove $PQRS$ is a rhombus, you must prove it's a parallelogram. Given that $\overline{PR}$ bisects $\angle SPQ$ and $\angle SRQ$, you can show that $\angle SPR \cong \angle QPR$ and that $\angle SRP \cong \angle QRP$. Because $\angle SPQ \cong \angle SRQ$ and halves of equals are equal, all four angles are congruent: $\angle SPR \cong \angle QPR \cong \angle SRP \cong \angle QRP$. Since $\angle SPR \cong \angle QRP$, $\overline{SP} \parallel \overline{RQ}$, and since $\angle QPR \cong \angle SRP$, $\overline{SR} \parallel \overline{PQ}$. Showing that both pairs of opposite sides are parallel proves that $PQRS$ is a parallelogram. The fact that $\angle SPR \cong \angle SRP$ ensures that $\overline{SP} \cong \overline{SR}$, and likewise $\angle QPR \cong \angle QRP$ means that $\overline{PQ} \cong \overline{QR}$, which is enough to prove $PQRS$ is a rhombus.
2. Since $ABDF$ is a rectangle, $\overline{AB} \perp \overline{BD}$ and $\overline{FD} \perp \overline{BD}$, so both ΔABC and ΔFDE are right triangles. Given $ACEF$ is a parallelogram and $ABDF$ is a rectangle, you know that $\overline{AC} \cong \overline{EF}$ and $\overline{AB} \cong \overline{DF}$, because opposite sides of any parallelogram (including a rectangle) are congruent. That's enough to prove $\Delta ABC \cong \Delta FDE$ by HL. Finally by CPCTC, $\overline{BC} \cong \overline{DE}$.
3. To prove $WXYZ$ is an isosceles trapezoid, you first have to prove it's a trapezoid. Since you are given $\overline{VX} \cong \overline{VY}$ and $\overline{WV} \cong \overline{VZ}$, work with the fact that ΔXVY and ΔWVZ are both isosceles triangles. The vertex angles of these triangles are congruent ($\angle XVY \cong \angle WVZ$ by vertical angles). That fact, with a little bit of arithmetic, should be enough to show that $\angle VXY$, $\angle VYX$, $\angle VWZ$, and $\angle VZW$ are all the same size. (Each triangle has angles that total 180°. You know the vertex angles are the same size, so what remains of the 180° is split between the base angles.) Since $\angle VXY \cong \angle VZW$, $\overline{XY} \parallel \overline{WZ}$ and $WXYZ$ is a trapezoid. Then to the given you could add that vertical angles $\angle XVW \cong \angle YVZ$, and prove $\Delta XVW \cong \Delta YVZ$, so $\overline{XW} \cong \overline{YZ}$ and $WXYZ$ is an isosceles trapezoid.
4. Because $\overline{RW} \perp \overline{TS}$ and $\overline{PV} \perp \overline{QU}$, ΔPVQ and ΔRWS are right triangles. Since $PQRS$ is a square, $\overline{PQ} \cong \overline{RS}$, and you're given that $\overline{RW} \cong \overline{PV}$. $\Delta PVQ \cong \Delta RWS$ by HL, and $\angle PQV \cong \angle RSW$ by CPCTC. Because a square is a rectangle, $\angle QPU$ and $\angle SRT$ are right angles, and all right angles are congruent. $\Delta QPU \cong \Delta SRT$ by ASA, and $\overline{PU} \cong \overline{TR}$. Opposite sides of the original square are parallel and congruent, so $\overline{QT} \parallel \overline{US}$, and by subtraction you can show that $QR - TR = PS - PU$, so $\overline{QT} \cong \overline{US}$. Since one pair of opposite sides is parallel and congruent, $QUST$ is a parallelogram.
5. Since $RTVW$ is a trapezoid, $\overline{TV} \parallel \overline{RW}$ and therefore $\overline{SV} \parallel \overline{RW}$. Prove $\Delta RTS \cong \Delta UTV$ by SAS, using the vertical angles, and you can conclude that $\angle SRT \cong \angle VUT$ by CPCTC. This means that $\overline{RS} \parallel \overline{WU}$. Since both pairs of opposite sides are parallel, $RSVW$ is a parallelogram.

7 Similarity

7·1

1. $x = 4.2$
2. $w = 15$
3. $x = 25$
4. $x = 216$
5. $x = 15$
6. $x = 29$
7. $x = \pm 8$
8. $x = \pm 5$
9. $x = \pm 13$
10. $x = 7, x = -4$

7·2

1. $\frac{AB}{VW} = \frac{BC}{WX} = \frac{CD}{XY} = \frac{DE}{YZ} = \frac{AE}{VZ}$
2. $\frac{ST}{MA} = \frac{TE}{AI} = \frac{EP}{IL} = \frac{SP}{ML}$
3. $\frac{CA}{DO} = \frac{AT}{OG} = \frac{CT}{DG}$
4. $\frac{HA}{FO} = \frac{AN}{OR} = \frac{ND}{RK} = \frac{HD}{FK}$
5. $\frac{BR}{MI} = \frac{RA}{IS} = \frac{AN}{ST} = \frac{NC}{TE} = \frac{CH}{ED} = \frac{BH}{MD}$

7·3

1. $\Delta WUV \sim \Delta YXZ$, AA
2. $\Delta ART \sim \Delta SUM$, AA (Because both triangles are equilateral, they are both equiangular; all angles are 60°.) *Or you could use SSS.*
3. Cannot be determined
4. $\Delta AGE \sim \Delta LDO$, SSS
5. $\Delta UER \sim \Delta PAO$, SAS
6. $\Delta BDA \sim \Delta CDB$, AA
7. Cannot be determined
8. $\Delta HOU \sim \Delta ESU$, AA (Because $\overline{HO} \parallel \overline{SE}$, alternate interior angles are congruent.)

9. Cannot be determined
10. $\Delta PHE \sim \Delta PON$, AA (Because $\overline{HE} \parallel \overline{ON}$, alternate interior angles are congruent.)

7·4

1. $\frac{AB}{XY} = \frac{BC}{YZ} = \frac{AC}{XZ}$
2. $\frac{RS}{FE} = \frac{RT}{ED} = \frac{RT}{FD}$
3. $\frac{PQ}{VX} = \frac{QR}{XW} = \frac{PR}{VW}$
4. $\frac{ML}{LJ} = \frac{LN}{JK} = \frac{MN}{LK}$
5. $\frac{ZX}{BC} = \frac{XY}{CA} = \frac{ZY}{BA}$
6. $\Delta ABC \sim \Delta XYZ$
7. $\Delta RST \sim \Delta OMN$
8. $\Delta PQR \sim \Delta EDF$
9. $\Delta ABC \sim \Delta RTS$
10. $\Delta XYZ \sim \Delta PRQ$
11. $YZ = 18$ cm
12. $AT = 9$ in
13. $AM = 10\frac{2}{3}$ ft
14. $DO = 20.4$ m
15. $LE = 14$
16. $TO = 42$
17. $EA = 23$
18. $LI = 49$
19. $CP = 3$
20. $AT = 3$

7·5

1. WZ
2. YZ
3. YX
4. YW
5. WZ
6. $WZ = 24$
7. $YV = 3$
8. $WZ = 18$
9. $XZ = 81$
10. $WZ = 12$ or $WZ = 54$ [$x = 4$ or $x = 18$]

7·6

1. $EF = 6$
2. $EF = 25.5$
3. $BC = 99$
4. $EF = 57$
5. $ED = 73.2$
6. $AC = 6$
7. $DE = 95$
8. $DF = 48$
9. $DE = 120$
10. $BC = 12$

7·7

1. Given $\overline{HO} \parallel \overline{SE}$, show that $\angle H \cong \angle E$ and $\angle O \cong \angle S$ because alternate interior angles are congruent. Prove $\Delta HOU \sim \Delta ESU$ by AA, and then $OU/US = HU/UE$ because corresponding sides of similar triangles are in proportion.
2. Given $\angle XWY \cong \angle XZV$ and $\angle X \cong \angle X$ by the reflective property, show $\Delta VXZ \sim \Delta YXW$ by AA. You can conclude $VX/YX = XZ/YW$ because corresponding sides of similar triangles are in proportion.
3. Given $\overline{VW} \parallel \overline{XZ}$, use the side-splitter theorem to show that $VY/VX = YW/WZ$, then cross-multiply to show $VY \cdot WZ = YW \cdot VX$.
4. Working backward from what you're asked to prove, $MN = \frac{1}{2}BD$, determine that you need to prove either $\Delta MAN \sim \Delta DCB$ or $\Delta MAN \sim \Delta BAD$. The proofs are similar. Let's do $\Delta MAN \sim \Delta BAD$. $\angle A \cong \angle A$ by the reflexive property. Since M is the midpoint of $\overline{AB}$ and N is the midpoint of $\overline{AD}$, $AM/AB = AN/AD = \frac{1}{2}$. By SAS, $\Delta MAN \sim \Delta BAD$. Since all the corresponding sides of similar triangles are in proportion, you know that $MN/BD = \frac{1}{2}$ as well. Multiply both sides by BD and $MN = \frac{1}{2}BD$.
5. Working backward from what you need to prove, $YZ \cdot XZ = VZ^2$, determine that you need to prove $\Delta VYZ \sim \Delta XVZ$. You're given $\angle X \cong \angle ZVY$ and you have $\angle Z \cong \angle Z$ by the reflexive property, so you can show $\Delta VYZ \sim \Delta XVZ$ by AA. The corresponding sides are in proportion, so $YZ/VZ = VZ/XZ$, and then cross-multiply to show $YZ \cdot XZ = VZ^2$.

8 Right triangles

8·1

1. $c = 5$
2. $b = 12$
3. $a = 24$
4. $c = \sqrt{130} \approx 11.4$
5. $b = \sqrt{5} \approx 2.2$
6. $c = \sqrt{5} \approx 2.2$
7. $c = 7\sqrt{2} \approx 9.9$
8. $b = 11\sqrt{3} \approx 19.1$
9. $b = 72$
10. $c = \sqrt{802} \approx 28.3$

8·2

1. Obtuse
2. Acute
3. Acute
4. Obtuse
5. Acute
6. Right
7. Obtuse
8. Right
9. Obtuse
10. Right

8·3

1. $BC=5, AC=10$
2. $AB=7, AC=14$
3. $BC=9, AB=9\sqrt{3}$
4. $AC=34, BC=17\sqrt{3}$
5. $AC=46, AB=23\sqrt{3}$
6. $AB=BC=8$
7. $BC=5, AC=5\sqrt{2}$
8. $AB=3, AC=3\sqrt{2}$
9. $AB=BC=6\sqrt{2}$
10. $BC=4\sqrt{2}, AC=8$
11. $AB=15, AC=10\sqrt{3}$
12. $AB=BC=9\sqrt{2}/2$
13. $AC=4\sqrt{5}, BC=2\sqrt{15}$
14. $AB=2\sqrt{3}, AC=2\sqrt{6}$
15. $BC=12.5, AB=12.5\sqrt{3}$
16. $BC=11.2, AC=11.2\sqrt{2}\approx 15.8$
17. $AB=4\sqrt{3}, AC=8\sqrt{3}$
18. $AB=BC=8\sqrt{3}$
19. $BC=19\sqrt{3}/3, AC=38\sqrt{3}/3$
20. $AB=11\sqrt{5}/2, BC=11\sqrt{15}/2$

8·4

1. $YW=3$
2. $YW=4\sqrt{3}$
3. $XW=9$
4. $YW=8$
5. $YW=5\sqrt{2}$
6. $YW=6$
7. $YW=10$
8. $XZ=3\frac{1}{3}$
9. $XW=25$
10. $YW=5$

8·5

1. $XZ=9$
2. $WZ=4$
3. $XW=6$
4. $XW=77$
5. $YZ=2$
6. $XZ=20$
7. $XW=4$
8. $WZ=6.75$
9. $XZ=10$
10. $XZ=9$

9 Circles

9·1

1. Secant
2. Diameter
3. Chord
4. Radius
5. Tangent
6. $\overleftrightarrow{JK}$
7. $\overleftrightarrow{EH}$
8. $\overline{OJ}$, $\overline{OK}$, or $\overline{CD}$
9. $\overline{FG}$
10. $\overleftrightarrow{LB}$

9·2

1. $m\angle O=48°$
2. $m\angle P=43°$
3. $m\angle O=40°$
4. $m\angle P=33°$
5. $m\widehat{NQ}=45°$
6. $m\widehat{NQ}=106°$
7. $m\angle P=14°$
8. $m\angle SPN=115°$
9. $m\angle RPQ=130°$
10. $m\angle SPQ=50°$
11. $m\angle RPN=55°$
12. $m\angle SPQ=56°$
13. $m\angle SPQ=49°$
14. $m\angle RPN=74°$
15. $x=7$
16. $t=12$
17. $y=19$
18. $p=31$
19. $m\angle NPQ=50.5°$
20. $m\angle RPN=83°$

9·3

1. $m\angle 1=79°$
2. $m\angle 4=97°$
3. $m\angle 3=101°$
4. $m\angle 2=94°$
5. $m\widehat{BD}=85°$
6. $m\widehat{AD}=93°$
7. $m\angle 1=86°$
8. $m\widehat{BC}=231°$
9. $m\widehat{AC}=93°$
10. $m\angle 4=70°$

9·4

1. $m\angle WZX=31°$
2. $m\angle VZW=36°$
3. $m\angle WZY=37°$
4. $m\angle VZY=105°$
5. $m\widehat{VRT}=43°$
6. $m\widehat{RT}=13°$
7. $x=7$
8. $x=19$
9. $m\angle VZY=80°$
10. $m\widehat{VRT}=83°$

9·5

1. $PR=12$
2. $PT=28$
3. $PO=13$
4. $OR=21$
5. $m\angle TPR=37°$
6. $TP=8$
7. $x=17$
8. $PR=12$
9. $x=26$
10. $PT=84$

9·6

1. $ED=4$
2. $EB=8$
3. $CE=3$ or 13
4. $x=7$
5. $x=\frac{1}{2}$

6. $VW = 29$
7. $ZW = 1$. Algebraically, $ZW = 9$ is also possible, but Figure 9.7 suggests $ZW < OV$.
8. $XY = 12$
9. $x = 15$
10. $x = 21$

9·7

1. $FC = 42$
2. $FD = 5$
3. $DC = 10$
4. $FA = 21$
5. $FE = 8$
6. $FC = 9$
7. $FC = 64$
8. $DC = 6$ or $DC = 18$
9. $FA = 12$
10. $FE = 8$

9·8

1. $r = 2$, $C(0, 0)$
2. $r = 4$, $C(0, 0)$
3. $r = 3$, $C(2, 5)$
4. $r = 5$, $C(7, -1)$
5. $r = 1$, $C(-3, -9)$
6. $(x-2)^2 + (y-7)^2 = 9$
7. $(x+3)^2 + (y-4)^2 = 6.25$
8. $(x-9)^2 + (y+7)^2 = 64$
9. $(x+6)^2 + (y+4)^2 = 13$
10. $x^2 + (y-2)^2 = 147$

10 Trigonometry

10·1

1. $\sin Z = \frac{3}{5}$
2. $\cos X = \frac{3}{5}$
3. $\tan X = \frac{4}{3}$
4. $\cos Z = \frac{4}{5}$
5. $\sin X = \frac{4}{5}$
6. $\tan R = \frac{12}{5}$
7. $\sin S = \frac{5}{13}$
8. $\cos R = \frac{5}{13}$
9. $\tan S = \frac{5}{12}$
10. $\sin R = \frac{12}{13}$
11. 0.6018
12. 1
13. 0.5
14. 0.2588
15. 8.1443

10·2

1. $a = 17.9$
2. $b = 406.3$
3. $a = 137.1$
4. $b = 52.5$
5. $a = 90.9$
6. $c = 84.9$
7. $a = 50.7$
8. $c = 69.9$
9. $b = 35.1$
10. $c = 43.8$
11. 23.1 ft
12. 437.3 ft
13. 89.0 in
14. 790.1 ft
15. 9 cm

10·3

1. $\angle R = 30°$
2. $\angle R = 55.2°$
3. $\angle R = 53.1°$
4. $\angle S = 68.2°$
5. $\angle S = 53.1°$
6. $\angle R = 19.5°$
7. $\angle R = 75.5°$
8. $\angle S = 68.3°$
9. $\angle S = 30°$
10. $\angle R = 45.6°$
11. 22.6°
12. 53.1°
13. 75.5°
14. 63.9°
15. 70.5°

10·4

1. $a^2 = b^2 + c^2 - 2bc\cos A$

$$\cos A = \frac{-1}{-56} \approx 0.0179$$

$$\angle A = \cos^{-1}(0.0179) \approx 88.98°$$

2. $b^2 = a^2 + c^2 - 2ac\cos B$

$$\cos B = \frac{-97}{-112} \approx 0.8661$$

$$\angle B = \cos^{-1}(0.8661) \approx 29.99°$$

or

$$\frac{a}{\sin A} = \frac{b}{\sin B}$$

$$\sin B = \frac{0.9998}{2} = 0.4999$$

$$\angle B \approx 29.99°$$

3. $\dfrac{20}{\sin 42°} = \dfrac{b}{\sin 30°}$

$$b = \frac{(0.5)(20)}{0.6691} \approx 14.94$$

4. $\dfrac{20}{\sin 42°} = \dfrac{c}{\sin 108°}$

$$c = \frac{20\sin 108°}{\sin 42°} \approx 28.43$$

5. $t^2 = r^2 + s^2 - 2rs\cos T$

$t^2 = 9^2 + 6^2 - 2\cdot 9\cdot 6\cos 165°$

$t \approx 14.88$

6. $\frac{13}{\sin 65°} = \frac{AB}{\sin 10°}$ and $\frac{13}{\sin 65°} = \frac{AC}{\sin 105°}$

$AB = \frac{13\sin 10°}{\sin 65°} \approx 2.4908$ $\quad AC = \frac{13\sin 105°}{\sin 65°} \approx 13.8552$

7. $\frac{8}{\sin 30°} = \frac{16}{\sin X}$ The triangle is a 30°-60°-90° right triangle.

$\sin X = \frac{16(0.5)}{8} = 1$

$\angle X = \sin^{-1}(1) = 90°$

8. $(XZ)^2 = 42^2 + 53^2 - 2\cdot 42\cdot 53\cos 19°$

$XZ \approx 19.06$

9. $c^2 = a^2 + b^2 - 2ab\cos C$

$c^2 = 325^2 + 450^2 - 2\cdot 325\cdot 450\cdot \cos 115°$

$c = \sqrt{431{,}740.8416} \approx 657$

10. First, find the angles in ΔRFC. $\angle R = 40°$ and $\angle C = 60°$, so $\angle F = 80°$. Then

$\frac{RF}{\sin 60°} = \frac{10}{\sin 80°}$ so $RF = \frac{10\sin 60°}{\sin 80°} \approx 8.79$.

10·5

1. $A = \frac{1}{2}\cdot 8\cdot 12\cdot \sin 30° = 24$

2. $A = 4\cdot 7\cdot \sin 48° \approx 20.8081$

3. $A = 14\cdot 14\cdot \sin 56° \approx 162.4914$

4. $A = \frac{1}{2}\cdot 18\cdot 27\cdot \sin 15° \approx 62.8930$

5. $A = 36\cdot 15\cdot \sin 125° \approx 442.3421$

6. $A = \frac{1}{2}ac\sin B$

$17 = \frac{1}{2}\cdot 4\cdot b\cdot \sin 30°$

$17 = 2b\cdot \frac{1}{2}$

$b = 17$

7. $A = ab\sin C$

$296 = 18^2\sin C$

$296 = 324\sin C$

$\sin C = \frac{296}{324} \approx 0.9136$

$\angle C = \sin^{-1}(0.9136) \approx 66°$

8. $A = \frac{1}{2}ab\sin C$

$49 = \frac{1}{2}x^2\sin 130°$

$x^2 = \frac{98}{\sin 130°} \approx 127.9299$

$x \approx 11.31$

9. $A = \frac{1}{2}\cdot 9^2\cdot \sin 60° \approx 35.0740$

10. $A = 40x\sin 70°$

$751.75 = 40x\sin 70°$

$x = \frac{751.75}{40\sin 70°} \approx 20$

11 Coordinate geometry

11·1

1. $3\sqrt{10} \approx 9.5$
2. $\sqrt{17} \approx 4.1$
3. $\sqrt{29} \approx 5.4$
4. $\sqrt{34} \approx 5.8$
5. 7
6. 4 or 10
7. −9 or 15
8. 1 or 15
9. −10 or 8
10. 4 or −4

11·2

1. (3.5, 5.5)
2. (−2, 4.5)
3. (−3, −2)
4. (4, 4)
5. (2, −3)
6. $x = 2$
7. $x = 7$
8. $y = 9$
9. $y = -5$
10. $x = 16$

11·3

1. $-\frac{3}{5}$
2. $-\frac{2}{3}$
3. 0
4. $\frac{1}{4}$
5. Undefined
6. $y = -2$
7. $x = 4$
8. $y = 4.5$
9. $x = -8$
10. $y = 3$

11·4

1.

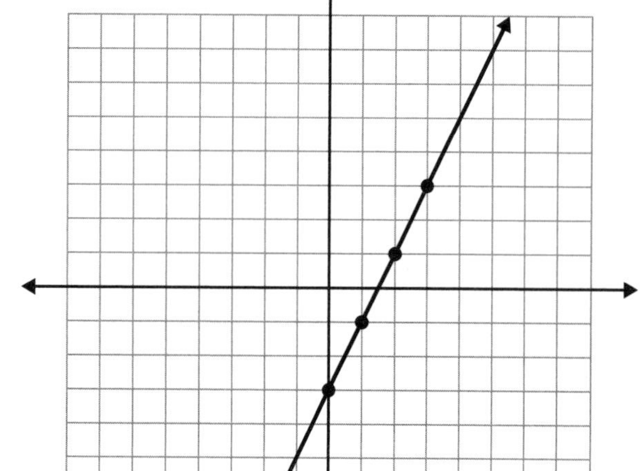

2.

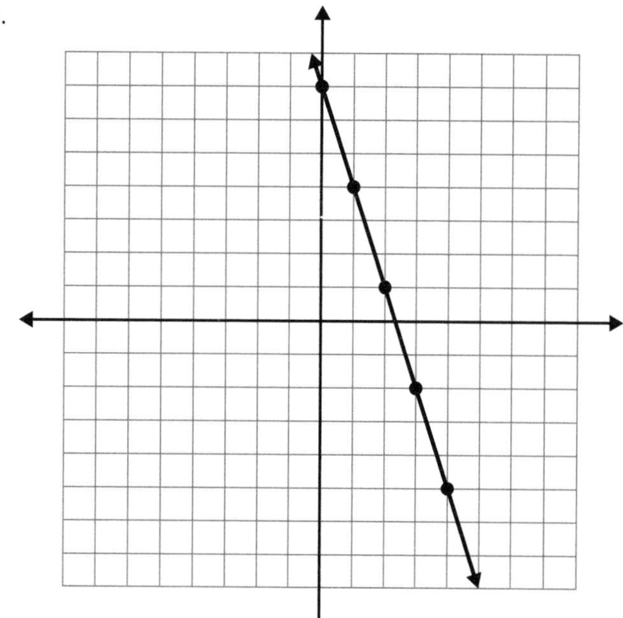

3.

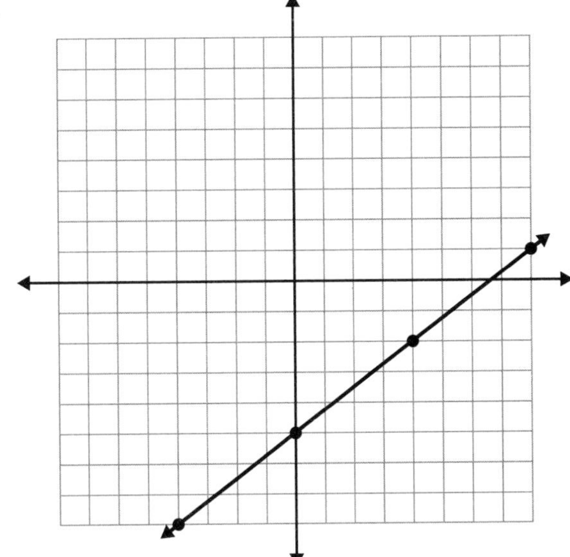

4.

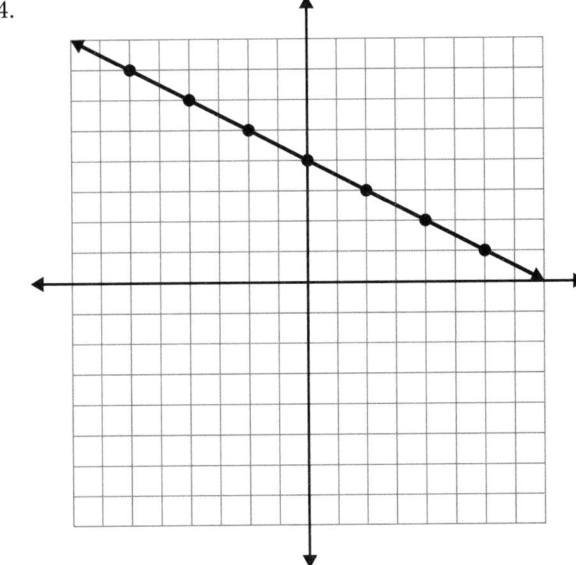

5.

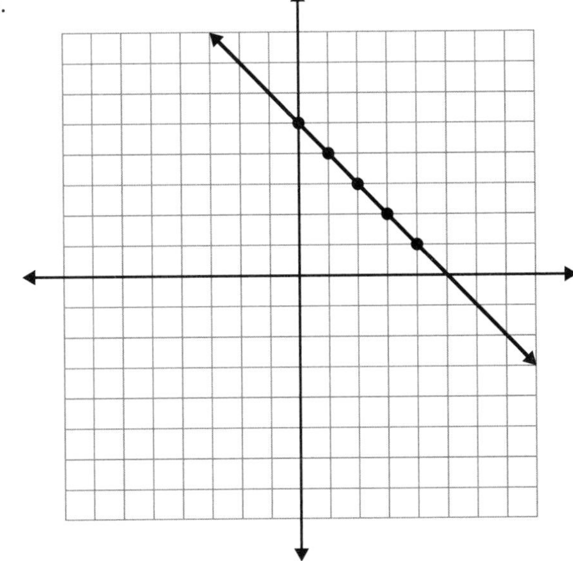

6.

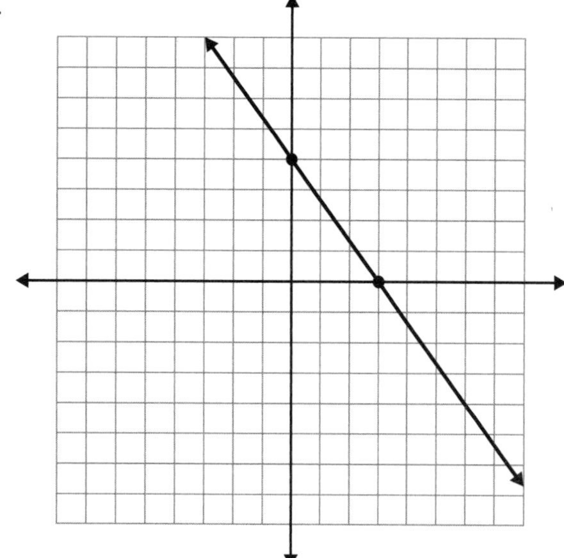

7.

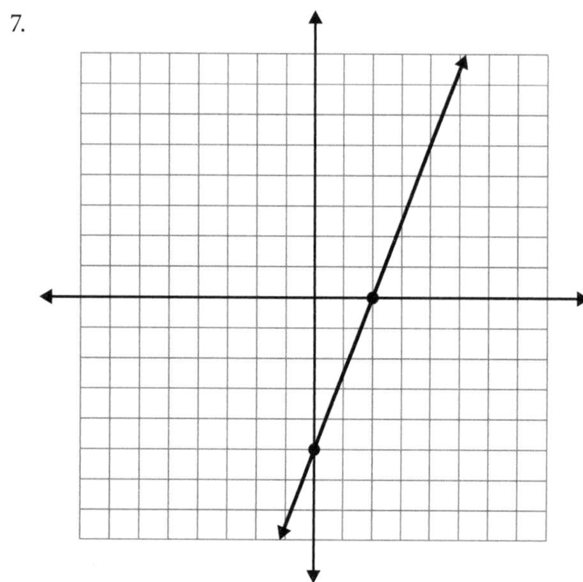

8.

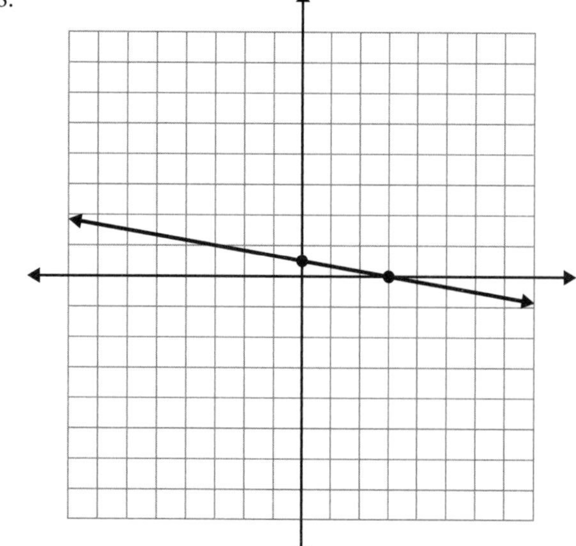

9.

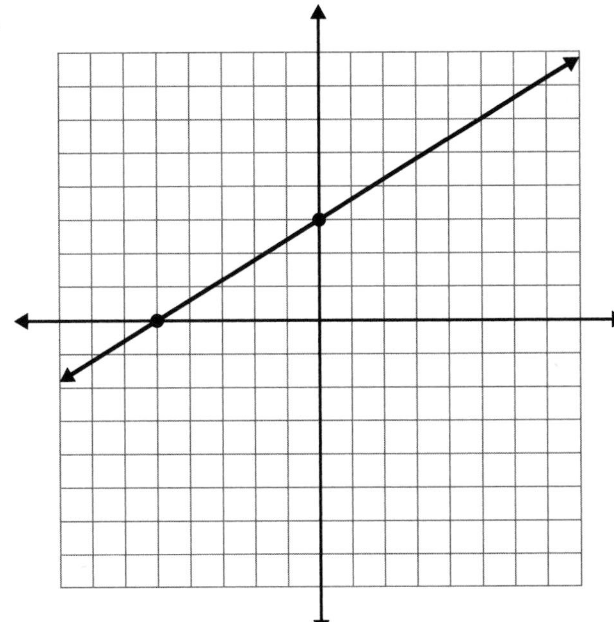

10.

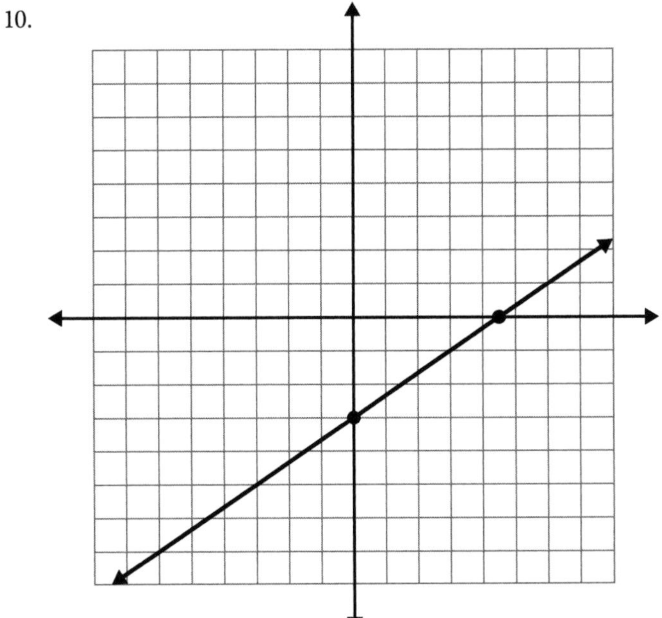

11·5

1. $3x+y=1$
2. $y=\frac{1}{3}x+2$
3. $2x+5y=5$
4. $y=\frac{5}{2}x+5$
5. $2x-y=-4$
6. $y=-x-1$
7. $x-y=-3$
8. $y=-\frac{1}{3}x+\frac{1}{3}$
9. $3x+4y=15$
10. $y=-\frac{9}{2}x+\frac{11}{2}$

11·6

1. Perpendicular
2. Parallel
3. Neither
4. Parallel
5. Perpendicular
6. $y=5x-16$
7. $y=-\frac{3}{4}x+2$
8. $y=-\frac{4}{3}x+\frac{7}{3}$
9. $y=-\frac{1}{4}x+14$
10. $y=2x-16$

11·7

1. The midpoint of one diagonal is $M_1 = \left(\frac{a+b+0}{2}, \frac{c+0}{2}\right) = \left(\frac{a+b}{2}, \frac{c}{2}\right)$ and the midpoint of the other diagonal is $M_2 = \left(\frac{a+b}{2}, \frac{0+c}{2}\right) = \left(\frac{a+b}{2}, \frac{c}{2}\right)$. Since the two diagonals have the same midpoint, they bisect each other.

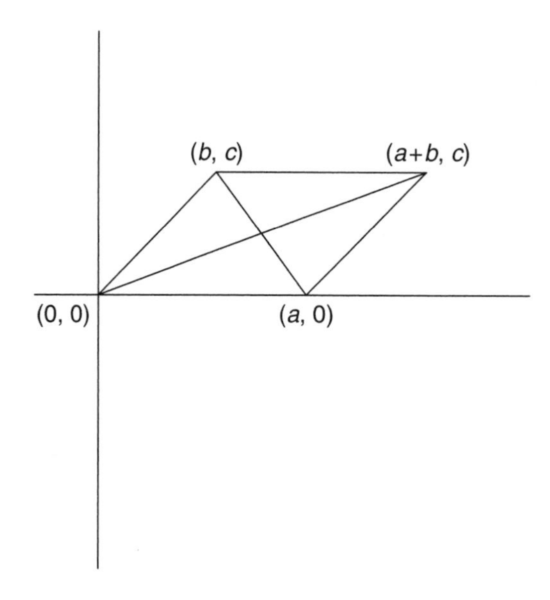

2. Find the lengths of the sides of the triangle. The length of the side to which the altitude is drawn is $2a$. The lengths of the other two sides are $\sqrt{(-a-0)^2+(0-b)^2} = \sqrt{a^2+b^2}$ and $\sqrt{(a-0)^2+(0-b)^2} = \sqrt{a^2+b^2}$. These two sides are the same length, so the triangle is isosceles.

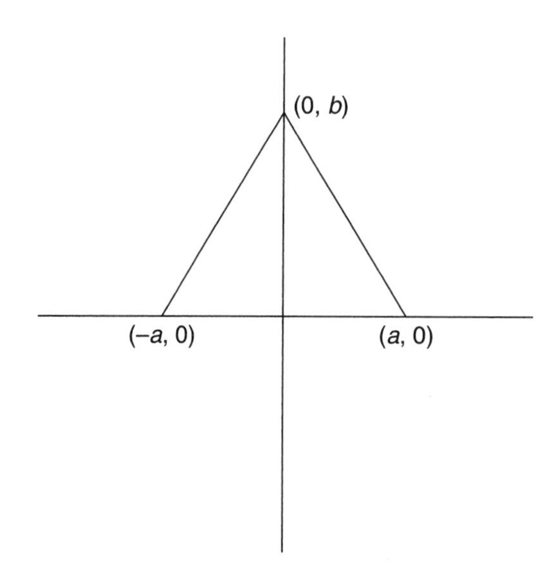

3. Find the lengths of the two diagonals. $d_1 = \sqrt{(a-0)^2+(b-0)^2} = \sqrt{a^2+b^2}$ and $d_2 = \sqrt{(a-0)^2+(0-b)^2} = \sqrt{a^2+b^2}$. The diagonals are congruent.

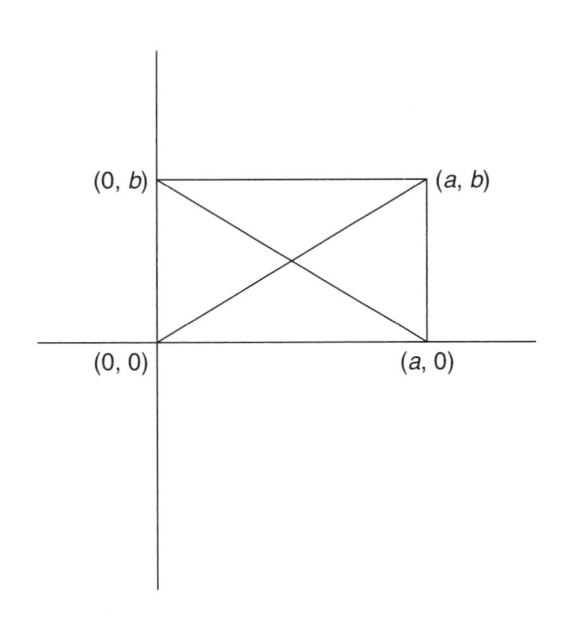

4. First, find the midpoints. $M_1 = \left(\frac{0+b}{2}, \frac{0+c}{2}\right) = \left(\frac{b}{2}, \frac{c}{2}\right)$ and $M_2 = \left(\frac{a+b}{2}, \frac{0+c}{2}\right) = \left(\frac{a+b}{2}, \frac{c}{2}\right)$. Next, show that the slope of the segment joining these midpoints is equal to the slope of the other side. The slope of the third side of the triangle is $m = \frac{0-0}{a-0} = 0$, and the slope of the segment connecting the midpoints is $m = \frac{c/2 - c/2}{(a+b)/2 - b/2} = 0$, so the segments are parallel. Finally, find the length of each, using the distance formula. $d_1 = \sqrt{(a-0)^2 + (0-0)^2} = a$ and $d_2 = \sqrt{\left(\frac{a+b}{2} - \frac{b}{2}\right)^2 + \left(\frac{c}{2} - \frac{c}{2}\right)^2} = \sqrt{\left(\frac{a}{2}\right)^2} = \frac{a}{2}$. Since $\frac{a}{2}$ is one-half of a, the segment joining the midpoint is one-half as long as the third side of the triangle.

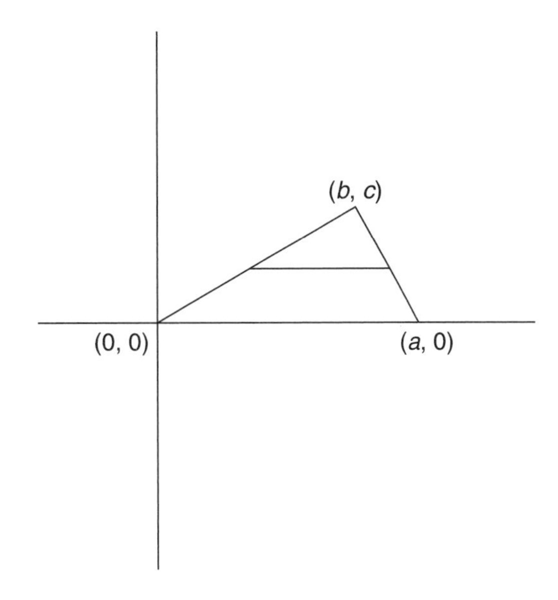

5. Find the slope of each diagonal. $m_1 = \frac{c-0}{a+b-0} = \frac{c}{a+b}$ and $m_2 = \frac{c-0}{b-a} = \frac{c}{b-a}$. To show that the diagonals are perpendicular, you must show that the slopes multiply to −1. $m_1 \cdot m_2 = \frac{c}{a+b} \cdot \frac{c}{b-a} = \frac{c^2}{b^2-a^2}$. Since $a^2 = b^2 + c^2$, substituting gives $m_1 \cdot m_2 = \frac{c^2}{b^2-a^2} = \frac{c^2}{b^2-(b^2+c^2)} = \frac{c^2}{-c^2} = -1$. The diagonals are perpendicular.

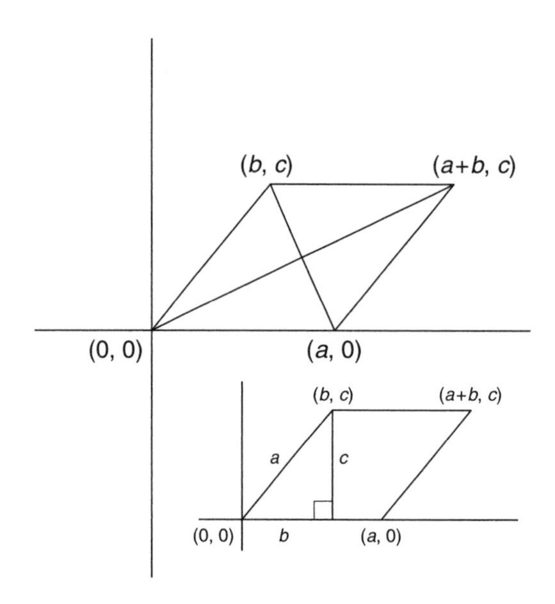

12 Transformations

12·1

1. $P'(2, 3)$
2. $Q'(1, 6)$
3. $R'(3, 4)$
4. $S'(-1, -5)$
5. $T'(0, 2)$
6. $A'(-2, -3)$, $B'(1, 7)$, $C'(-4, 3)$
7. $A'(4, -5)$, $B'(-3, -2)$, $C'(2, 3)$
8. $A'(7, 1)$, $B'(5, -2)$, $C'(-4, -3)$, $D'(-2, 3)$
9. $A'(-1, -1)$, $B'(7, -1)$, $C'(7, 2)$, $D'(-1, 2)$
10. $A'(-1, 4)$, $B'(-3, 1)$, $C'(-2, -4)$, $D'(2, -3)$, $E'(1, 3)$

12·2

1. $P'(8, 5)$
2. $P'(-1, 6)$
3. $P'(-1, 6)$
4. $P'(-8, 0)$
5. $P'(3, -4)$
6. $P(3, -6)$
7. $P(3, 3)$
8. $P(-7, 3)$
9. $P(10, -6)$
10. $P(-9, 11)$
11. $A'(1, 2)$, $B'(-2, 12)$, $C'(3, 8)$
12. $A'(7, 3)$, $B'(0, 0)$, $C'(5, -5)$
13. $A'(3, 10)$, $B'(0, 8)$, $C'(-1, -1)$, $D'(5, 1)$
14. $A'(-6, -3)$, $B'(2, -3)$, $C'(2, -6)$, $D'(-6, -6)$
15. $A'(5, 3)$, $B'(7, 0)$, $C'(6, -5)$, $D'(2, -4)$, $E(3, 2)$

12·3

1. (1, 4)
2. (2, 3)
3. (−1, 0)
4. (−4, 0)
5. (1, −3)
6. $A'(-3, -2)$, $B'(7, 1)$, $C'(3, -4)$
7. $A'(5, -4)$, $B'(2, 3)$, $C'(-3, -2)$
8. $A'(-7, 1)$, $B'(-5, -2)$, $C'(4, -3)$, $D'(2, 3)$
9. $A'(1, -1)$, $B'(-7, -1)$, $C'(-7, 2)$, $D'(1, 2)$
10. $A'(-4, 1)$, $B'(-1, 3)$, $C'(4, 2)$, $D'(3, -2)$, $E'(-3, -1)$

12·4

1. $A'(4, -6)$, $B'(-2, 14)$, $C'(8, 6)$
2. $A'(2, 2.5)$, $B'(-1.5,1)$, $C'(1, -1.5)$
3. $A'(1.5, 10.5)$, $B'(-3, 7.5)$, $C'(-4.5, -6)$, $D'(4.5, -3)$
4. $A'(-0.8, 0.8)$, $B'(5.6, 0.8)$, $C'(5.6, -1.6)$, $D'(-0.8, -1.6)$
5. $A'(3, 12)$, $B'(9, 3)$, $C'(6, -12)$, $D'(-6, -9)$, $E'(-3, 9)$
6. $A'(2, -3)$, $B'(-4, 17)$, $C'(6, 9)$
7. $A'(5, 9)$, $B'(-5.5, 4.5)$, $C'(2, -3)$
8. $A'(2, 2.5)$, $B'(0.5, 1.5)$, $C'(0, -3)$, $D'(3, -2)$
9. $A'(-1.8, 1.1)$, $B'(7, 1.1)$, $C'(7, -2.2)$, $D'(-1.8, -2.2)$
10. $A'(1, 4)$, $B'(7, -5)$, $C'(4, -20)$, $D'(-8, -17)$, $E'(-5, 1)$

13 Area, perimeter, and circumference

13·1

1. 112 cm^2
2. 81 m^2
3. 60 cm^2
4. 36 cm^2
5. 11 cm
6. 16 cm
7. 16 cm
8. 78 cm^2
9. 593 cm^2
10. 44 cm^2

13·2

1. 216 cm^2
2. 25 m^2
3. 112 cm^2
4. 513 cm^2
5. $150\sqrt{3}$ cm^2
6. 20 cm
7. 52 cm
8. 19 cm
9. 25 cm
10. $ZY = 23$ cm

13·3

1. 7.5 m^2
2. 1,743 cm^2
3. $32\sqrt{3}$ cm^2
4. 588 cm^2
5. $64\sqrt{3}$ cm^2
6. 13 cm
7. $7\sqrt{13}$ in $\approx$ 25.2 in
8. 21 m
9. 18 cm^2
10. $4\sqrt{6}$ cm $\approx$ 9.8 cm

13·4

1. $294\sqrt{3}$ cm^2
2. 480 cm^2
3. 300 cm^2
4. $190\sqrt{39} \approx$ 1,186.5 cm^2
5. $12\sqrt{3}$ cm^2
6. 27 cm
7. 24 cm
8. 65 cm
9. 18 cm
10. 12 cm

13·5

1. 25π cm^2
2. 121π m^2
3. 81π cm^2
4. 7 m
5. 22 cm
6. $18\sqrt{2\pi}$ in $\approx$ 80 in
7. \$209.23
8. 11π square units
9. 21.125π square units
10. 36π cm^2
11. 9 cm
12. 1.4π cm
13. 24 units
14. 24π cm
15. $46{,}656\pi$ cm^2

13·6

1. 428.75 m^2
2. 131.4 cm^2
3. 35 in
4. 272 m
5. 171 ft
6. 180 m^2
7. 1,183 ft^2
8. 23 cm
9. 2,849 cm^2
10. 4.5 in

13·7

1. You have a circle of diameter 12 in, inscribed in a square 12 in on a side. The area of the square is 144 in^2. The unshaded portion is $300°/360° = \frac{5}{6}$ of the area of the circle. The area of the circle is 36π in^2, so $\frac{5}{6} \times 36\pi = 30\pi$ is unshaded. The total shaded area is $144 - 30\pi$, or approximately 49.75 in^2.

2. This is a case when it's much simpler to calculate the shaded area directly. You have three shaded triangles, each with an area of $\frac{1}{2}\cdot 10\cdot 10 = 50$ cm², so the total shaded area is 150 cm².
3. The area of the white trapezoid is $\frac{1}{2}h\left(\frac{1}{2}b+\frac{1}{3}b\right)=\frac{1}{2}h\left(\frac{5}{6}b\right)=\frac{5}{12}bh$. Five-twelfths of the parallelogram is shaded.
4. The area of the outermost ring is the area of a circle of radius 9 minus the area of a circle of radius 7, or $81\pi - 49\pi = 32\pi$. The area of the inner shaded ring is the area of a circle of radius 5 minus the area of a circle of radius 3, or $25\pi - 9\pi = 16\pi$. The total shaded area is $32\pi + 16\pi = 48\pi$.

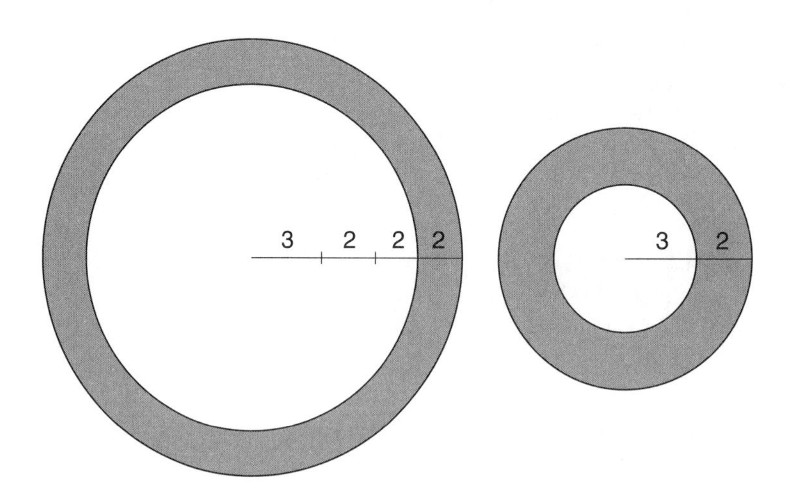

5. Drawing diagonals between opposite vertices (two are already drawn) divided the hexagon into six equilateral triangles, each 8 cm on a side. The shaded area is the area of the circle, radius 8 cm, minus the unshaded area, which is the equivalent of four of six equilateral triangles. The area of the circle is 64π cm². The area of one equilateral triangle is $\frac{1}{2}\cdot 8\cdot 4\sqrt{3} = 16\sqrt{3}$ cm², so the area of four such triangles is $64\sqrt{3}$ cm². The shaded area is $64\pi - 64\sqrt{3} \approx 90.21$ cm².
6. Each shaded rectangle will measure 12 cm by 6 cm and so will have an area of 72 cm². Four rectangles have a total area of 288 cm². Or you can make a rearrangement as in the following diagram and see that the shaded area is one-half the area of the outer square. $A = \frac{1}{2}\cdot 24^2 = 288$ cm².

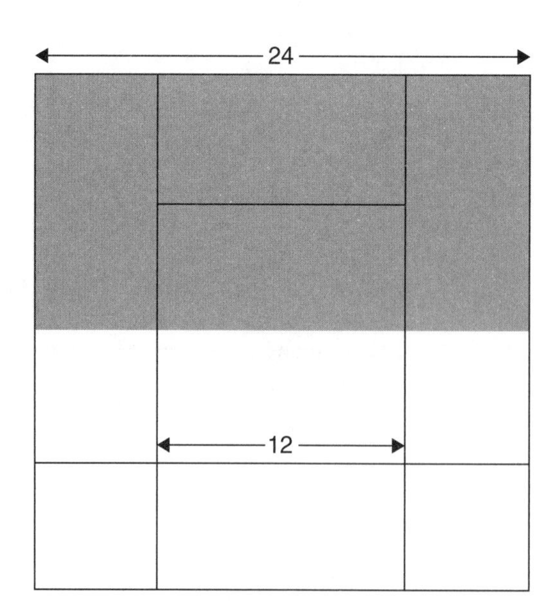

7. Decompose the figure into a rectangle and a ring. The area of the ring is the area of a circle of diameter 24 and therefore radius 12, minus the area of a circle of diameter 12 and radius 6. The area of the ring is $144\pi - 36\pi = 108\pi$. The area of the rectangle is $24 \times 4 = 96$. The total shaded area is $108\pi + 96$, or approximately 435.29 cm^2.

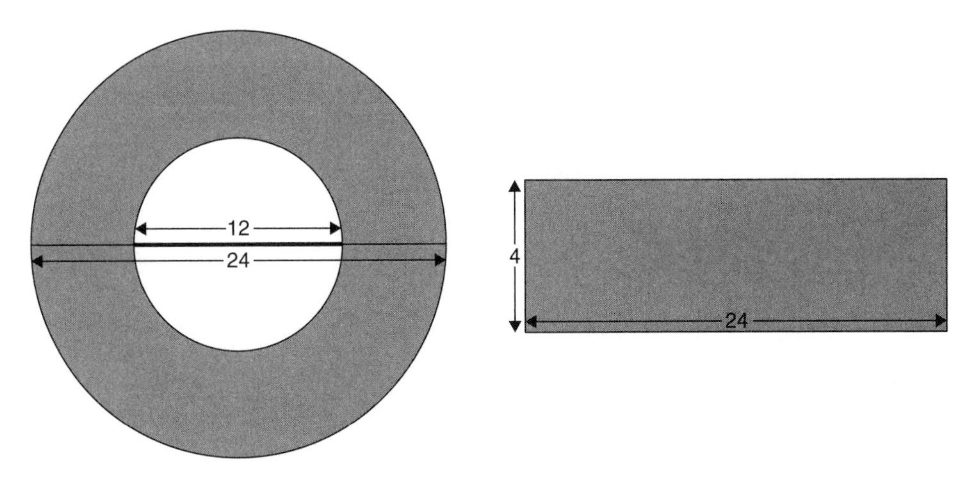

8. The picture looks fancy, but one-half the circle is shaded, so the shaded area is $\frac{1}{2}\pi \cdot 8^2 = 32\pi$ in^2.
9. Use the law of sines to find x in the shaded triangle. $15/\sin 108° = x/\sin 36°$ so $x = 15\sin 36°/\sin 108° \approx 9.27$. Use the Pythagorean theorem in the unshaded triangle to find h. Now $h^2 + (2.5)^2 = (9.27)^2$ so $h^2 = 85.9329 - 6.25 = 79.6829$ and $h \approx 8.27$. In the larger regular pentagon, use right-triangle trigonometry to find a. $\tan 36° = 7.5/a$ so $a = 7.5/\tan 36° \approx 10.32$. Area of the larger pentagon is $5 \cdot \frac{1}{2} \cdot 5 \cdot (10.32) \approx 129$. The area of one unshaded triangle is $\frac{1}{2} \cdot 5 \cdot (8.27) \approx 20.675$. The shaded area is area of the large pentagon minus total area of the five unshaded triangles, or $129 - 5(20.675) \approx 25.625$ cm^2.

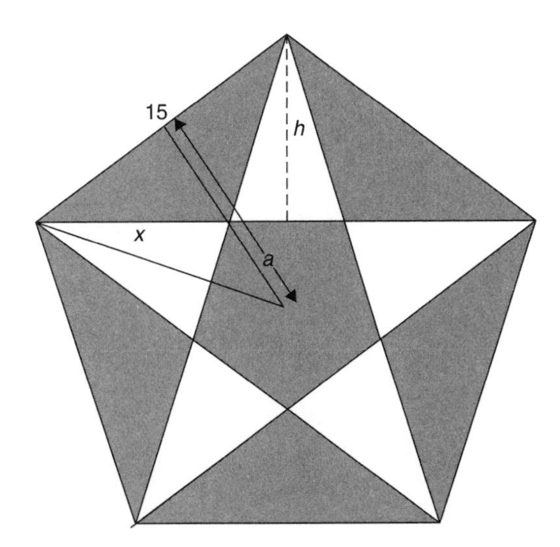

10. The area of the circle is 49π. Each unshaded triangle has a base of 2 and a height of 6, since the 14-cm diameter is reduced by 2 cm for the center square, leaving 12 cm to divide between the two heights. So each triangle has an area of $\frac{1}{2} \cdot 2 \cdot 6 = 6$ cm. The shaded area is the area of the circle minus four triangles, or $49\pi - 4 \cdot 6 = 49\pi - 24 \approx 129.94$ cm^2.

14 Surface area

14·1

1. $96 + 24\sqrt{10} \approx 171.9$ in^2
2. $288 + 192\sqrt{3} \approx 620.6$ cm^2
3. $96 + 8\sqrt{3} \approx 109.9$ cm^2
4. 687.5 in^2
5. 27 m^2
6. 8 cm
7. 7 cm
8. $\sqrt{3}$ cm
9. 40 cm
10. 18 cm

14·2
1. 1008π cm^2
2. 288π in^2
3. 104π in^2
4. 1634π cm^2
5. 120π in^2
6. 13 in
7. 6 m
8. 13 cm
9. 23 ft
10. 10 m

14·3
1. 10 in
2. $8\sqrt{3}$ cm
3. 37 ft
4. 16 m^2
5. $900\sqrt{3}$ in^2
6. $25+25\sqrt{3}$ cm^2
7. 414 in^2
8. 12 in
9. 26 cm
10. 315 cm^2

14·4
1. 17.2 cm
2. 21.5 in
3. 39 cm
4. 24π ft^2
5. 826π cm^2
6. $4{,}136\pi$ in^2
7. $20{,}250\pi$ cm^2
8. 19 cm
9. 2 in
10. 8 cm

14·5
1. 36π ft^2
2. 484π in^2
3. 144π m^2
4. 1,296 cm^2
5. 900π in^2
6. 6 ft
7. 25 cm
8. 19 cm
9. 8 m
10. 3 in

14·6
1. 1,075 cm^2
2. 1,584 cm^2
3. 1,147.5 cm^2
4. $4\sqrt{3}$ cm
5. $2\sqrt{2}$ cm
6. 2,511 cm^2
7. 2 in
8. 54π cm
9. 2 in
10. 10 cm

15 Volume

15·1
1. 16 in^3
2. 144 in^3
3. $576\sqrt{3}$ cm^3
4. $32\sqrt{3}$ cm^3
5. 990 in^3
6. 9 m^3
7. 8 cm
8. 7 cm
9. 17 cm
10. 8 cm

15·2
1. 108π cm^3
2. $4{,}320\pi$ cm^3
3. 567π in^3
4. 144 in^3
5. $8{,}664\pi$ cm^3
6. 13 in
7. 6 m
8. 13 cm
9. 23 ft
10. 20 m

15·3
1. $54\sqrt{3}$ cm^3
2. 4 m^3
3. $2{,}200\sqrt{3}$ in^3
4. $100\sqrt{3}$ cm^3
5. 367.5 in^3
6. $\frac{8}{3}\sqrt{3}$ ft^3
7. 30 in
8. 12 cm
9. 257 cm^2
10. $6\sqrt{3}$ in

15·4
1. 864π cm^3
2. 12π ft^3
3. $2{,}940\pi$ cm^3
4. $29{,}400\pi$ in^3
5. $337{,}500\pi$ cm^3
6. 750π cm^3
7. 15 cm
8. $3\sqrt{3}$ in
9. 9 cm
10. 142 in^3

15·5
1. 972π in^3
2. 36π ft^3
3. $1{,}774\frac{2}{3}\pi$ in^3
4. 288π m^3
5. $7{,}776\pi$ cm^3
6. $4{,}500\pi$ in^3
7. 3 in
8. 6 in
9. $7{,}812\pi$ in^3
10. $10{,}666\frac{2}{3}$ ft^3

15·6
1. 1,000 cm^3
2. 35 cm^3
3. 34,425 cm^3
4. 7 cm
5. 3 cm
6. 22,599 cm^3
7. 325 cm^3
8. 54π cm
9. 2 in
10. 10 cm